Thank you for your purchase of *Understanding Today's Electricity Business*. If you wish to purchase additional copies of this book, please visit our website at www.enerdynamics.com. Or call us at 866.765.5432. Volume discounts start at as few as twenty-five books.

Please look for the newest addition to our line of books: *Understanding Today's Global LNG Business*. Please also look for this book's natural gas companion, *Understanding Today's Natural Gas Business*. As with our electric book, this presents a comprehensive overview of the natural gas industry in simple and easy-to-understand language. Priced at $64.95, it is the perfect primer for those new and not-so-new to the industry, and a valuable reference for years to come.

We also invite you to experience other learning opportunities available from Enerdynamics. These include public and in-house seminars, self-paced online training, as well as our free *Energy Insider* newsletter. Learn more about all these products at www.enerdynamics.com.

Understanding Today's Electricity Business

By Bob Shively and John Ferrare

3101 Kintzley Court, Suite F
Laporte, CO 80535
866.765.5432
www.enerdynamics.com

Enerdynamics Corp.

Enerdynamics is an education firm dedicated to preparing energy industry employees for success in a challenging environment. We offer an array of public and in-house educational opportunities including classroom seminars, online seminars, and books. We can be contacted at 866-765-5432 or info@enerdynamics.com.

Please visit our website at www.enerdynamics.com

About the Authors

Bob Shively has over twenty years of experience in the gas and electric industries. As a partner in Enerdynamics, Bob has advised and educated some of the largest energy industry participants on issues ranging from market strategies to industry restructuring. Bob has also served as Vice-President of eServices of Sixth Dimension, Inc., an energy networking company, where he worked closely with retail marketing and ESCO companies in development of new economic demand response and distributed generation products. Bob began his career in the energy industry at Pacific Gas and Electric Company (PG&E). At PG&E Bob held various positions including Major Account Executive to some of PG&E's largest end-use customers and Director of Gas Services Marketing where he was responsible for product development and sales for the company's $1.5 billion dollar Canadian pipeline project. Bob has Master of Science degrees in both Mechanical and Civil Engineering from Stanford University and is a frequent energy industry speaker

John Ferrare has worked in the energy industry as a marketing and communications specialist for over fifteen years. He began his career with Pacific Gas and Electric Company where he was integral in developing the marketing group for the company's Gas Services Marketing Department. Since that time, he has also worked with PG&E Corporation and PG&E Energy Services in the development of marketing and communication strategies. In 1995, John joined Enerdynamics to manage its educational services. In this role, he has helped create a comprehensive program to educate 600 utility employees on the changes brought by recent deregulation as well as the core classes currently offered by Enerdynamics. A graduate of Northwestern University's School of Speech, John has also developed and teaches a public speaking and communications class for a variety of corporate audiences.

ISBN 978-0-9741744-1-9

Edition 5.0

Copyright ©2010. All rights reserved. No part of this book may be reproduced or transmitted in any form by any means, electronic, mechanical, photocopying, recording, or otherwise, without the prior written permission of Enerdynamics Corp.

While every precaution has been taken in the development of this book, Enerdynamics Corp. makes no warranty as to the accuracy or appropriateness of this material for any purpose. Enerdynamics Corp. shall have no liability to any person or entity with respect to any loss or damage caused or alleged to be caused directly or indirectly by material presented in this book.

The authors of this book wish to thank Dan Bihn, Chuck Sathrum, Karen Shea, and Greg Stark for the immeasurable improvements they suggested in reviewing drafts. We wish to thank the analysts at the Energy Information Administration whose data is used throughout our book. And Tim Collins whose extraordinary illustrations often tell so much more than our words ever could.

John would also like to recognize Jesse, Cody, Rudy, and Athena, whose collective barks, hugs, licks, and genuine concern and affection offered considerable motivation during the long hours of editing this document.

Bob thanks Carol for her support and love, and thanks Jed and Tarah for understanding that his computer is necessary for something besides playing online games.

And finally, we wish to thank the thousands of participants in Enerdynamics' seminars and programs, who in the last fifteen years have taught the authors more than we could ever have imagined.

CONTENTS

SECTION ONE: INTRODUCTION 1
   Today's Electricity Marketplace 1
   Electricity in Modern Society 4
   A Brief History of Electricity 5

SECTION TWO: WHAT IS ELECTRICITY? 9
   How Electricity is Created 11
   How Electricity is Used to Perform Useful Tasks 12
   The Key Components of the Electric Delivery System 15
   The Key Physical Properties of an Electric Delivery System 16
      Electricity Cannot be Stored 16
      The Path of Electrical Flow is Difficult to Control 16
      Disturbances Travel Very Quickly and are Hard to Contain 17
      Outages and Significant Voltage or Frequency Fluctuations are Not Acceptable ...17
   The Four Key Physical Sectors of the Electricity Business 18

SECTION THREE: ELECTRIC CONSUMERS 21
   Residential Customers 22
   Commercial Customers 25
   Industrial Customers 28
   Aggregate Demand Curves 32

SECTION FOUR: GENERATION 35
   Types of Generation 36
      Coal 36
      Nuclear 38
      Natural Gas 38
      Hydro 39
      Fuel Oil 40
      Renewables 41
      Distributed Generation 42

CONTENTS

    Environmental Considerations 43
    Electric Generation, Global Warming and the Kyoto Protocol 44
    Demand Response as an Alternative to Generation 46
    Use of Generation to Satisfy the Load Curve 47
        Baseload 48
        Intermediate 48
        Peaking 48
    Ownership of Generation 49
    Developing a Generation Portfolio 50
    The Future of Generation 51

SECTION FIVE: TRANSMISSION 55
    Physical Characteristics of Transmission 56
    Operation and Planning of the Transmission System 57
    Transmission System Costs 58
    Ownership of Transmission 58
    Issues with Transmission Construction 59
    The Current Status of the U.S. Transmission System 60

SECTION SIX: DISTRIBUTION 63
    Physical Characteristics of Distribution 63
        Radial Feed 67
        Loop Feed 67
        Network System 67
    Operating and Planning of the Distribution System 68
    Distribution System Costs 69
    Ownership and the Current Status of Distribution Systems 69
    The Smart Grid 70

SECTION SEVEN: ELECTRIC SYSTEM OPERATIONS 73
    Operational Characteristics of Power Systems 73
    What System Operations Does 74
    Who Handles System Operations 74
    Forecasting and Scheduling 77
        Demand Forecasting 78
        Scheduling Generation, Transmission and Reserves 78
    Ancillary Services 78
        Automatic Generation Control (AGC) 79
        Spinning Reserves 79
        Non-spinning Reserves 80

      Supplemental Reserves 80
      Voltage Support 80
      Black Start 80
   How Supply and Demand are Kept in Balance in Real Time 80
   The Changing Role of System Operations 82

SECTION EIGHT: MARKET PARTICIPANTS IN THE DELIVERY CHAIN 85
   Participants in the Vertically-Integrated Market Model 85
      Investor-Owned Utilities 85
      Municipal Utilities and Public Utility Districts 86
      Rural Electric Co-ops 86
      Federal Power Agencies 87
      Public Power Agencies 88
      Power Pools 88
      Energy Services Companies (ESCOs) 88
      Independent Power Producers and Electric Marketers 88
   Participants in Restructured or Competitive Electric Markets 89
      Merchant Generators 89
      Transmission Companies 89
      Independent System Operators (ISOs)/Regional Transmission Organizations (RTOs) .90
      Electric Marketers 90
      Financial Services Companies 91
      Transmission Owners 91
      Utility Distribution Companies 91
      Load Serving Entities 92
      Energy Services Companies (ESCOs) 92

SECTION NINE: ELECTRIC MARKET STRUCTURES 95
   What is an Electric Market Structure? 95
   Vertically-Integrated Monopoly Utility Model 97
      The Current Status of Vertically-Integrated Monopoly Utility 98
   Single Buyer with Competitive Generation Model 99
      The Current Status of Single Buyer 100
   Wholesale/Industrial Competition Model 100
      The Current Status of Wholesale/Industrial Competition 102
   Complete Retail Competition Model 102
      The Current Status of Complete Retail Competition 103
   Trading Arrangements 105
      Wheeling 105

CONTENTS

      Decentralized 106

      Integrated 107

   The Current Status of Market Structures and Trading Arrangements 109

SECTION TEN: REGULATION IN THE ELECTRIC INDUSTRY 111

   Why Regulate the Electric Industry? 111

   The Goals of Regulators 112

   The Historical Basis for Regulation 112

      State Regulation 112

      Federal Regulation 113

   Who Regulates What? 115

   The Regulatory Process 116

      The Initial Filing 116

      Preliminary Procedures 117

      Hearings 117

      The Draft Decision 118

      The Final Decision 118

      Review of Decisions 118

      Tariffs 118

   Setting Rates through a Traditional Ratecase 119

      Determining the Authorized Rate of Return .120

      Forecasting Usage 120

      Determining a Revenue Requirement 120

      Allocating Revenue to Customer Classes 121

      Determining Rate Design 121

      Allocating Revenue to Charge Types 122

      Determining the Rate 122

   Incentive Regulation 122

      Performance-based 123

      Benchmarking 123

      Rate Caps 123

      Service Standards 123

   Market-based Rates 124

   State and Federal Rate Methodologies 124

   The Future of Regulation 124

SECTION ELEVEN: THE CONCEPTS OF MARKET RESTRUCTURING 127

   Why Restructuring? 128

   Market Evolution under Deregulation 129

Regulation 130
Deregulation 131
Commoditization 132
Value-Added Services 132
The Necessary Components for a Competitive Marketplace 133
Supply Side Competition 133
Fair Access to Transmission 134
Unbiased System Operations 134
Demand Side Competition 135
Distribution without Impediments to Competition 136
Opportunities for Hedging Risks 136
Creating a Competitive Market 136
Transitioning Generation 137
Creating a Robust Transmission Market 138
Creating an ISO 138
Transitioning to Customer Choice 139
Continued Regulation of Transmission and Distribution 139
Ensuring Reliability 139
Settlements 140

SECTION TWELVE: THE HISTORY OF ELECTRIC MARKET RESTRUCTURING ..143
Federal Restructuring 144
The First Steps Towards Independent Generation – PURPA of 1978 144
Fostering Wholesale Generation Competition – the Energy Policy Act of 1992 . .144
Furthering Open Access Transmission – FERC Order 888 146
Encouraging Formation of Regional Transmission Operators – FERC Order 2000 . .148
The Current Stalemate 149
State Restructuring 150
Separating the Vertical Utility Functions 152
Allowing Retail Access 152
Continued Regulation of the Monopoly Function 152
The California Experience 153
Restructuring in Other Countries 156

SECTION THIRTEEN: MARKET DYNAMICS 159
Supply and Demand 160
Short-term Supply and Demand 161
Long-term Supply and Demand 162

CONTENTS

       The Current Supply/Demand Situation in the U.S. 162
   Pricing 163
       Indexes and Trading Hubs 164
       Price Volatility 165
   The Wholesale Market 166
       Energy and Generation Capacity 168
           Forward Markets 168
           Spot Markets 169
       Transmission Rights 169
       Financial Services 170
   The Retail Market 170
       Utility Retail Services 171
       Competitive Retail Services 172
       ESCO and other Energy Services 173

SECTION FOURTEEN: MAKING MONEY AND MANAGING RISK 175
   How Market Participants Create Profits 176
       How a Utility Makes Money —Traditional Method 176
       How a Utility Makes Money — Incentive Regulation 178
       How Unregulated Market Participants Make Money 178
   Risk Management 178
       Choices for Managing Risk 180
           Physical Risk Management 180
           Financial Risk Management 181
           Speculation versus Hedging 182
           Hedging Techniques 182
       Value at Risk 184

SECTION FIFTEEN: THE FUTURE OF THE ELECTRICITY BUSINESS 187
   A Review of Market Changes 187
   The Future of the Generation Sector 188
   The Future of Transmission 189
   The Future of Distribution 189
   The Future of System Operations 190
   The Future of Retail Marketing 190
   A Sustainable Energy Future? 191

APPENDIX A: GLOSSARY 195

APPENDIX B: UNITS AND CONVERSIONS 207

APPENDIX C: ACRONYMS...209

APPENDIX D: INDEX 215

What you will learn:

- An overview of today's electricity marketplace
- The importance of electricity in today's society
- Electric usage levels in the U.S. and the world
- A brief history of the electrical industry

SECTION ONE: INTRODUCTION

Today's Electricity Marketplace

A world with an insatiable appetite for electricity awakens to a new era. Technological advances rapidly expand the possibilities of what electricity can do and how it can be created. Consumers demand access to a continually advancing array of products and services based on the ready availability of reliable electric supply. In some regions, regulated utilities dominate the market and are the only electric providers available to consumers. In others, all customers are free to choose their electric supplier and service providers compete vigorously. Government regulation, once the stable backbone of the industry, struggles to keep pace with marketplace developments and the conflicting demands of varied market participants. And concerns over environmental impacts create a push to change traditional generation sources. Meanwhile, rapidly developing societies across the globe clamor for modern conveniences, causing worldwide demand to double every 25 years. This is today's electric marketplace.

It's hard to imagine, but just over one hundred years ago Thomas Edison was drawing up the papers to create the Edison Electric Light Company. At that time, the electric lights and motors that Edison would power had yet to be invented. Now our society would scarcely exist as we know it without the ubiquitous movement of electrons known as electricity. As we move ever deeper into the information age, virtually all our activities are dependent on one commodity – electricity.

And paralleling our increasing dependence on electricity is an electric marketplace marked by rapid change. Market structures that remained stable for close to one hundred years are being radically transformed. The staid vertically-integrated utility is, in many areas of the world, being forced to separate into distinct generation, transmission, distribution, and retail services companies. Generation and retail services are being opened to competition. In some areas, customers who never thought twice about their electric service now shop for retail suppliers as casually as they do long distance providers. And caught in the wake of this turbulent industry restructuring are the

SECTION ONE: INTRODUCTION

fates of many corporate enterprises, some whose future prosperity suddenly seems to be in question.

The technology of electricity continues to evolve as well. In the 1990s, the improvement of gas turbine generation technologies enabled competitive generating units to provide electricity more cost-effectively than the vertical utility. Next, wind generation became a viable technology and created opportunities to diversify our generation portfolio with significant amounts of renewable power. And recent technology innovations promise a continued moving target. Possible futures range from a centralized supply based on clean coal or refined nuclear technologies to a distributed supply based on small non-polluting solar cells located at customers' facilities. Superconductivity may allow us to greatly enhance our current transmission system without building new towers. And smart grid technology that would allow distribution systems to continually communicate with customers and enable new services could revolutionize the distribution industry. The possibilities, it would seem, are endless.

In this book, we'll take an in-depth look at this rapidly evolving industry. We'll begin with an overview of U.S. and global electricity use and some history on how we've gotten to where we are today. Next we'll look at what electricity is and how it gets to consumers. Then a look at who electricity consumers are and how they use electricity. From there we'll examine in depth the three physical sectors of the delivery system – generation, transmission and distribution – and how these sectors are operated. We will study electric market structures and explore who does what and how these market participants are organized and interact. Then we'll open the doors to regulation and deregulation and take a look at how the electric industry has evolved to where we are today. Next we'll study the dynamics of the market and how participants attempt to make money and manage risk. And finally, we'll sneak a peak into the future and speculate on what exciting changes may lie ahead.

As you may know, the electricity business is filled with acronyms and industry-specific jargon which will be important for you to understand. For this reason, we suggest you begin your study with a look at the glossary and the list of acronyms found at the end of this book. We also suggest you refer back to the glossary whenever you find a word you don't understand. Once you are comfortable with these terms, feel free to study the information contained in this book in any order that makes sense for you. Good luck and have fun!

A WORD ABOUT UNITS

We cannot effectively discuss electricity without referring to the units by which it is measured. So it's extremely important that you clearly understand the following concepts.

Units of demand/capacity

Demand reflects the instantaneous amount of work required to perform the function desired (e.g., creating light or physical force, powering a microchip, etc.). Similarly, capacity reflects the instantaneous ability to provide energy required to do work (e.g., generation and transmission capacity, etc.). Demand and capacity are measured in units of watts, kilowatts, megawatts, or gigawatts:

$$1 \text{ kilowatt (kW)} = 1000 \text{ watts}$$

$$1 \text{ megawatt (MW)} = 1000 \text{ kW}$$

$$1 \text{ gigawatt (GW)} = 1000 \text{ MW}$$

As an example, consider a 60 watt light bulb in your living room lamp. When you turn the lamp on, the bulb creates a demand of 60 watts. And in order for this electricity to be supplied to the bulb, 60 watts of capacity must be available at the generator and along the entire path between the bulb and the generation source.

Units of energy/usage

Energy or usage reflects demand or capacity multiplied by the amount of time that demand or capacity is in use. Energy and usage are measured in units of watt-hours, kilowatt-hours, megawatt-hours, or gigawatt-hours.

$$1 \text{ kilowatt-hour (kWh)} = 1000 \text{ watt-hours}$$

$$1 \text{ megawatt-hour (MWh)} = 1000 \text{ kWh}$$

$$1 \text{ gigawatt-hour (GWh)} = 1000 \text{ MWh}$$

Let's consider again our living room lamp. The 60 watt bulb it uses to light your living room uses 60 watt-hours of energy if you leave it on for one hour. If you leave it on for a 24-hour period, it would consume 1440 watt-hours (60 watts X 24 hours), which is the same as 1.44 kWh.

Of course, that's just for one bulb. An average suburban home in the United States (with a single air-conditioning unit) has a peak demand of about 3-5 kW, with an average demand of approximately 1 kW. Peak demand refers to the greatest amount of electricity required at any given moment. Typical average monthly usage for this home would be about 720 kWh. Total U.S. consumption is approximately 4 trillion kWh per year, which breaks down to about 13,000 kWh per person. Peak demand in the U.S. is approximately 760,000 MW.

SECTION ONE: INTRODUCTION

Electricity in Modern Society

For much of the last century, increases in the United States' consumption of energy have closely matched growth of the U.S. economy. In the last half century, electric use has failed to grow from year-to-year only five times (in 1974, 1982, 2001, 2008, and 2009). Per-capita average consumption in the U.S. has also grown considerably. While the U.S. population grew by 89% over the past 50 years, electricity use grew by an astonishing 1,315%!

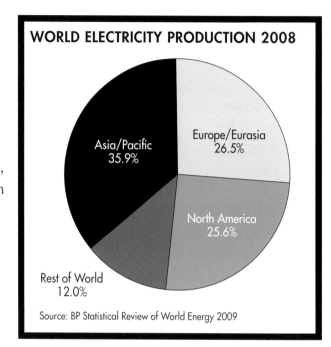

Electricity use continues to grow around the world as well. In 2007, the world consumed over 17 trillion kWh[1] which was more than double the consumption in 1975. This growth is not expected to abate anytime soon. Unbelievably, over two billion people currently lack access to electricity. If we assume globalization continues to bring affluence to developing regions, many of these will expect to share our modern standards of living and will soon contribute to even greater growth in world electric demand.

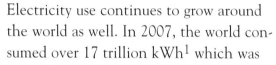

According to the Energy Information Administration (EIA), worldwide electricity production is projected to increase at an average annual rate of 2.4% over the next twenty-five years. While use in the industrialized world is expected to increase at a historically modest rate of 1.2% per year, use in the developing countries is expected to increase by 3.5% annually. By the year 2030, developing counties are expected to increase their percentage

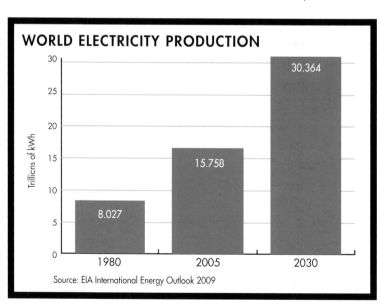

---

[1] The U.S. alone was responsible for almost 23% of world consumption.

of world usage from today's 45% to 58%. At this time, world electricity output will have almost doubled from output in the year 2005, repeating the near doubling of output in the twenty-five years between 1980 and 2005.

Of course, our hunger for electric power is not without significant cost. Consumers in the United States spent about $365 billion on electricity in 2008. The International Energy Agency has estimated that the world will need to spend $12 trillion over the next 30 years just to provide the necessary electric infrastructure that will accommodate growing demand. At the same time, electric generation – fueled worldwide by the burning of hydrocarbons – continues to negatively impact our natural environment. As we move into the 21st century, many observers now agree that mankind must develop new, more environmentally benign, ways of generating and using electricity to create a sustainable future.

A Brief History of Electricity

The word electricity is derived from the Latin word electricus which means to "produce from amber by friction." As long ago as 600 B.C., Greeks knew that amber could be charged by rubbing. And it was Thales of Miletus who many credit with the first discovery of electricity. He noticed that rubbing two pieces of amber together created a force that could attract light objects such as cat fur.

While man has long known that the phenomenon we now call electricity existed, it was not until much later that it was studied and ultimately put to use. In 1600, the English scientist William Gilbert described the electrification of many different substances and coined the word electricity. By 1660, scientists had invented a machine for producing static electricity by rotating and rubbing a ball of sulfur.

As society entered the 1700s, scientists began the work of harnessing electricity. In 1729, the conduction of electricity was discovered and subsequent work identified substances that would act as conductors. And in 1745, the Leyden jar was invented. This device contained a glass vial partially filled with water and a thick wire that could conduct an electrical charge. The Leyden jar was notable in that it was the first device that allowed electric charge to be stored and later discharged all at once (leading ultimately to the concept of electric current). It was also during this time that Benjamin Franklin invented the lightning rod, proving that lightning was actually a form of electricity.

In 1799, copper and zinc plates separated by cardboard soaked in salt water were used to create the first continuous and controlled source of electricity, the forerunner of our

SECTION ONE: INTRODUCTION

modern day battery. Having a steady source of electricity pushed researchers' capabilities into a new realm, and by the mid 1800s we had laws that described the basic behavior of electricity (Ohm's law and Kirchoff's law) as well as various electric technologies such as the electromagnet, the electric motor, the electric generator, and electric arc lights. We also saw one of the first practical uses of electricity – the telegraph, invented by Samuel Morse around 1840.

As the world entered the industrial age, scientists and engineers pressed to create additional practical uses of electricity. The discovery of the self-excited dynamo – a generator that could quickly ramp up to full capacity – made it feasible to create small scale generating stations, and by the mid-1870s electric arcs were lighting the streets of Paris, London and New York. Arc lights, however, were too powerful for indoor uses and widespread use of electricity awaited Thomas Edison's development of a practical incandescent bulb in 1879. Soon after this discovery, Edison patented his design for an electrical distribution system. On September 4, 1882, the Edison Electric Illuminating Company opened its first central generating station at Pearl Street in Manhattan, and the era of the electric utility was born. Soon thereafter, a number of small distribution systems were created and Benjamin Harrison, elected in 1888, became the first president to have electricity in the White House (although, as the story goes, he and his wife were afraid of being shocked by the light switches, so they opted to use the gas lights instead). Use of electricity to power street railways also became common, and by 1889 there were 154 of them in the U.S.

Broader development of electric utilities was hamstrung by the fact that initial systems were direct current (DC). DC systems at voltages safe for home use cannot efficiently distribute power for more than about a half mile due to losses of electricity along the distribution wires. At that time, raising voltages to transmit power and then lowering them again for household use was too expensive to be feasible. The breakthrough that led to today's modern utility came in 1888, when Nikola Tesla created workable alternating current (AC) generators and motors. Unlike DC power, AC power can be transmitted longer distances at low voltages without undue losses. Using Tesla's patents, George Westinghouse received a contract to construct a power plant at Niagara Falls. The plant opened in August 1895 and powered two 3.7 MW generators. Initially the power was used locally for the manufacture of aluminum and carborundum, but in 1896 a 20-mile transmission line was constructed to Buffalo where the power was used for lighting and street cars.

Soon the electrical giants Westinghouse and General Electric came to dominate electric power technology, and in the years between 1900 and the First World War new electrical appliances such as the refrigerator, washing machine, vacuum cleaner, and radiant heating as well as improved light bulbs led to increasing demand for electricity. The concept of the vertically-integrated utility which owned and operated electric generation, transmission and distribution soon dominated the industry, leading to the need for regulation and, in some areas, municipalization of the utility function. Meanwhile, as part of the Depression era New Deal, the federal government began construction of numerous federal hydro power projects.

Between 1945 and 1965 the utility industry continued to grow and investors became familiar with the steady, if unspectacular, returns provided by utilities. The technology for central generating units powered by coal, fuel oil and natural gas matured, providing for increased efficiency and lower costs. Customers remained generally satisfied as technological innovations boosted transmission and distribution reliability at the same time that rates were falling.

1965 marked the start of a new era. The Northeast blackout in November 1965 left over 30,000,000 customers – including all of New York City – without power. This was perhaps our first realization that we had become dependent on an interconnected system that was less reliable than assumed. As we entered the 1970s, many utilities became enamored with the potential of nuclear energy, and construction began on a number of large nuclear generating units. Shocks soon to follow included the Arab Oil Embargo of 1973-1974, the Three Mile Island nuclear accident in March 1979 and subsequent rapidly increasing generation costs for many utilities. By the 1980s, many utilities were burdened with high debt levels and interest rates, incomplete power plant projects, slowing growth in the demand for electricity, the need for substantial electric rate increases, and increasing environmental concerns. Suddenly utilities found themselves portrayed in a negative light.

In the 1990s and early 2000s, the turmoil continued. Encouraged by deregulation in natural gas, airlines, transportation, and some foreign electric industries, free market advocates began pushing for competition in the electric industry. This led to the breakup of many vertically-integrated utilities, the spectacular rise and fall of marketing companies such as Enron and Dynegy, financial difficulties driven by expansion of utilities into unregulated activities, and a U.S. marketplace split into varying market structures. Today's marketplace continues to evolve as energy companies strive for success amid concerns over service reliability, the need for new generation sources and global warming.

What you will learn:

- What electricity is
- What an electrical current is
- How electricity is created
- How electricity is used to perform useful tasks
- The physical delivery system
- Key physical properties of electricity
- The four key physical sectors of the electric business

SECTION TWO: WHAT IS ELECTRICITY?

Electricity is simply the flow of electrons through a conductor[1]. Electrons are the tiny negatively charged particles that are found in all atoms. Electricity is transmitted as loose electrons move from one atom to the next within a conductor. A conductor is any material that facilitates this transmittal of electricity.

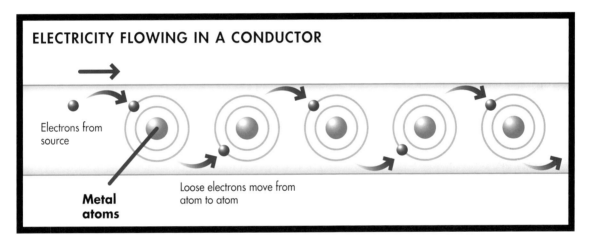

Most practical applications of electricity require electrons to flow through a circuit. A circuit includes a source of electrons (a battery or generator), an energy consuming device (such as a light bulb), and conductors (wire) that transmit the electrons to and from the bulb. In the simple circuit illustrated on page 10, the battery causes electrons to flow through the wire to the light bulb where light is created. The electrons then return to the battery via the wire and the electric circuit is complete. Note that the bulb does not "consume" the electrons, but rather the electrons flow through a material in the bulb causing it to glow.

Before we continue, there are several quantitative terms associated with the flow of electrons you will need to understand. The rate at which electrons flow through a conductor is called current and is measured by amperes or amps (A). If we were to compare the flow of electrons through a conductor to the flow of water through a hose,

---

[1] Throughout this book we will use the term electricity to mean the flow of electrons through a conductor, which is also known as current electricity. There is a second type of electricity known as static electricity. Static electricity is the transfer of electrons from one material to another and is not discussed in this book.

SECTION TWO: WHAT IS ELECTRICITY?

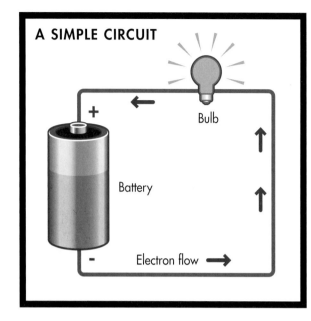

this rate would be the equivalent of gallons per second. There are two factors that affect this rate – the force applied to the electrons and the resistance to flow within the conductor. The force that moves electrons is called voltage, and is measured in volts (V). The higher the voltage, the faster the rate of flow (the higher the current). Voltage, then, is the equivalent of pressure in our hose. Resistance, which is measured in ohms (R), impedes the flow of electrons. As you might imagine, the higher the resistance in the conductor, the slower the rate of flow (the lower the current). Returning to our water analogy, ohms are equivalent to any friction or blockage that might slow down the flow of water through the hose. Current is always directly proportional to voltage and resistance as shown in a relationship known as Ohm's law:

$$\text{Current (amps)} = \frac{\text{Voltage (volts)}}{\text{Resistance (ohms)}}$$

SIX TYPES OF ENERGY

Electrical energy is one of six major types of energy:

Chemical energy — Stored energy released as the result of two or more atoms and/or molecules combining to form a chemical compound.

Electrical energy — Energy associated with the flow of electrons.

Electromagnetic energy — Energy associated with electromagnetic radiation including visible light, infra-red light, ultraviolet light, x-rays, microwaves, radio waves, and gamma rays.

Mechanical energy — Energy that can be used to raise a weight.

Nuclear energy — Stored energy released as the result of particles interacting with or within an atomic nucleus.

Thermal energy — Energy associated with atomic and molecular vibrations that results in heat.

Electrical energy stands alone in value among the six because it can be easily transported and readily transformed into other useful energy forms: to mechanical energy via an electric motor; to electromagnetic energy via light bulbs, microwave ovens, etc.; and to thermal radiation via radiant heaters.

This means that you would maintain the same current in a circuit if you increased both the volts and the ohms by the same ratio. You would increase the current if you increased the voltage but did not change the ohms. Conversely, you would decrease the current if you left the voltage unchanged, but increased the ohms.

Different voltages are used in a circuit depending on what is being done with the electricity. High voltages are used to transmit electricity long distances while lower voltages are used to power home appliances and office equipment. Voltages can be changed through the use of transformers. Transformers are able to change voltages because applying electricity to two different sized coils of wire in near proximity results in a voltage transformation. By adjusting the coil size, engineers can adjust voltage as required.

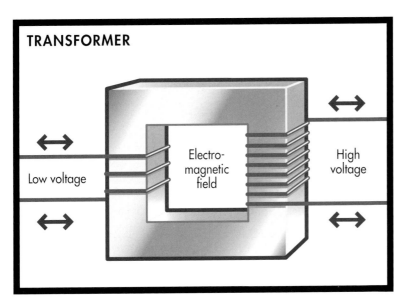

Most home appliances in the U.S. are operated on 110 V. Large appliances such as electric dryers and some electric ovens operate on a higher voltage of 220 V. We commonly refer to home electric services from the utility as 120/240 V services. Notice that while voltage may be supplied to a home at 120 V, the appliances use 110 V. The difference in voltage is due to resistance in home wiring. Commercial and industrial buildings often operate a number of their devices at higher voltages such 480 V. The size of an electrical service is determined in amps. A typical home built today in the U.S. would have a service of 200 amps.

How Electricity is Created

To begin the flow of electrons through a conductor, a source of energy is required. This can be either chemical or electromagnetic energy. Batteries and fuel cells operate by using chemical energy to free electrons from one material and transfer them to another via a conductor. Batteries and fuel cells contain three components – two electrodes and an electrolyte. The electrolyte reacts with the electrodes to create oxides that result in excess negative charge in one electrode (creating the negative terminal) and excess positive charge on the other (creating the positive terminal). The imbalance in charge creates an electric current when the terminals are connected to form a circuit.

Electromagnetic energy is used in two primary ways to create electricity. Solar or photovoltaic (PV) cells are made of materials that cause electrons to flow when light strikes the cell. As with a battery, the flow is directed through a circuit. The most common way of creating electricity – the electric generator – uses electromagnetic energy in a very different way. An electric generator creates electricity by what is called electromagnetic induction. Electromagnetic induction uses magnetism to make electrons flow. A source of mechanical energy (a steam turbine, gas turbine, wind turbine, or water turbine) is used to spin a shaft connected to a coil. This coil is suspended between the poles of a magnet and is connected to wires in a circuit by metallic brushes. As the coil spins through the magnetic field, electrons flow through the coil and brushes and then into the electric circuit.

> **ELECTRICAL TERMINOLOGY**
>
> **Current** — The rate of flow of electrons.
>
> **Amps** — The unit used to measure current.
>
> **Voltage** — The force that moves electrons.
>
> **Volts** — The unit used to measure voltage.
>
> **Kilovolts** — Another unit used to measure voltage, equal to 1000 volts.
>
> **Resistance** — A measure of the strength of impedance, which is the physical property that slows down electrons.
>
> **Ohms** — The unit used to measure resistance.
>
> **Transformer** — A device used to change voltages in a circuit.

How Electricity is Used to Perform Useful Tasks

You now know that electricity is the flow of electrons through a conductor. You also know that it flows through a conductor in a circuit under variable rates of flow. If that was all electricity did, however, it's unlikely you'd be reading about it in this book! What we've yet to discuss is that moving electrons create specific effects that can be harnessed to perform useful tasks. These effects include magnetism, heat and light. To understand electricity it is important to remember that unlike, say natural gas, electrons are not consumed while creating value. They can, however, be directed through specific materials resulting in all kinds of useful by-products. For example, light can be created by moving electrons through a filament of tungsten wire that is wound in a tight coil. The electrical flow causes the coil to heat up so that it becomes white hot and glows. This is the principle behind the incandescent light bulb, which is used commonly in our homes. The fluorescent light bulb works on a different principle. When an electrical current moves through certain gasses, they emit ultraviolet light.

UNDERSTANDING TODAY'S ELECTRICITY BUSINESS

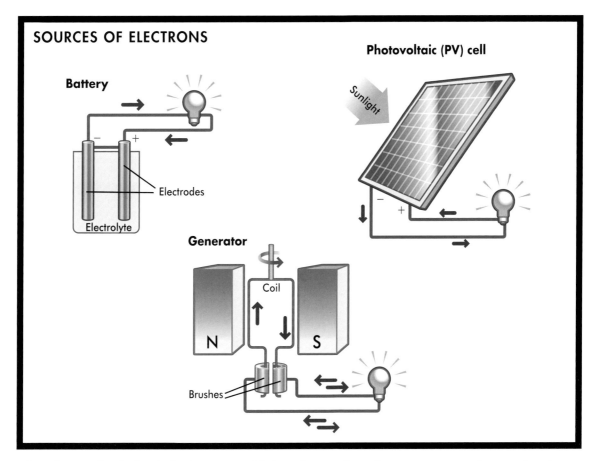

The ultraviolet light (which is invisible) strikes a phosphor coating on the inside of the fluorescent tube, causing the phosphor atoms to glow. Electric motors are made possible by the fact that moving electrons create magnetic fields. When properly harnessed, these magnetic fields can be used to spin a shaft creating mechanical energy which can then be used in a variety of ways.

One additional way that electricity is used to create value involves the principle of control. By controlling the flow of electrons, electronic devices can be used to transfer information. This is the principle behind transistors and microchips, the two devices that have made modern technology possible[2]. For example, by controlling the flow in a specific way, we can represent the number 1 or 0, allowing us to digitize information.

The amount of electricity necessary to perform useful tasks is determined by the power required to move the electrons through a specific device. This power is measured in units called watts. A typical incandescent light bulb requires 100 watts to operate. A

---

[2] By far, the best explanation of how electrical devices work in plain language is a book found on many a child's bookshelf titled *The New Way Things Work*, by David Macaulay.

SECTION TWO: WHAT IS ELECTRICITY?

fluorescent bulb, on the other hand, requires only 40 watts to deliver comparable light. (Thus the push to replace incandescent bulbs with fluorescent ones!) Power used over time is commonly called energy. So anyone using that 100 watt bulb for one hour will consume 100 watt-hours of energy.

The current, voltage and power available in a circuit are related:

$$1 \text{ watt} = 1 \text{ volt} \times 1 \text{ amp}$$

So the power available in a home service of 200 A delivered at 120 V is:

$$\text{Power (watts)} = 120 \text{ V} \times 200 \text{ A} = 24,000 \text{ W (which is the same as 24 kW)}$$

As you have learned, electricity is delivered to the devices that use it via an electrical circuit. If a circuit is not complete, the flow of electrons will stop. Thus, in all circuits – including the electrical distribution system – electrons flow from one end of the source, through the conductor and devices, and then back to the source.

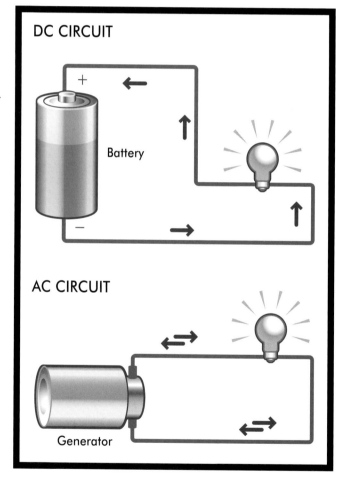

Now that you understand how current flows through a circuit, we need to confuse matters just a bit. Electric currents come in two types – direct current (DC) and alternating current (AC). In DC circuits the power flows continually in one direction, from the negative terminal to the positive terminal. In an AC circuit, the direction in which the electrons flow changes periodically and repeatedly. Electrons first flow from the generator towards the load. The flow then reverses direction and again flows to the load but from the opposite direction. Each time the direction of flow changes we say that the electricity has completed one-half cycle. A full cycle, then, is when the electricity flows first in one direction, and then the opposite. The unit of hertz (hz) is used to measure frequency, or how often electrons change direction. Utilities in the U.S. operate on a standard of 60 hz (meaning that electrons complete 60 cycles per second). Unfortunately, this is not the standard elsewhere in the world. Europe operates on 50 hz, which explains why your hair dryer requires a converter in your London hotel.

The Key Components of the Electric Delivery System

Now let's take what you've learned about a simple circuit and expand it to explain how electricity is created and delivered to consumers. An electric delivery system is, in its basic sense, simply a very large circuit. The flow of electrons, or current, is created by the generator. The electrons are transmitted to and back from consumers via conductors – transmission and distribution lines. And completing the circuit are the millions of energy-consuming devices.

The voltage created by generators is generally several thousand volts. This voltage is then stepped-up to transmission voltage by a step-up transformer. Banks of step-up transformers are typically located in a substation immediately adjacent to the genera-

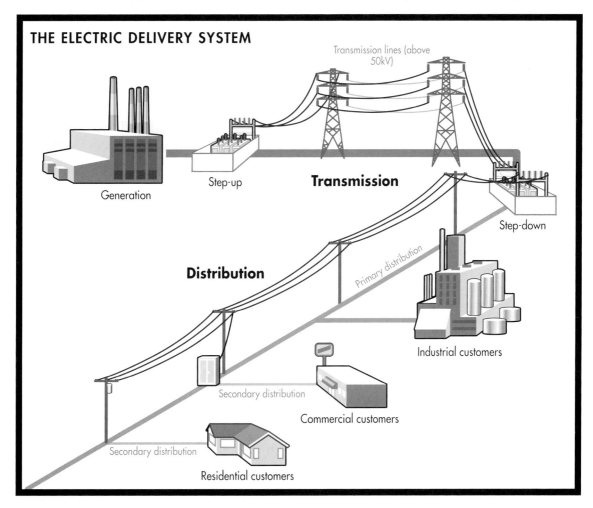

tor. These transformers facilitate electric transmission because it is much more efficient to move electricity long distances at high voltages. Unfortunately, current at high voltages is capable of sparking or jumping large distances and is extremely dangerous to humans. This is why high voltage transmission lines are located on large towers.

As the electricity approaches an area where it will be consumed, the voltage is dropped to a safer voltage. This is performed by a distribution transformer located in a distribution substation. The electricity then continues its journey via the lower voltage distribution lines until it reaches the service line (the line entering a consumer's building). At the interconnection with the service line, the voltage is often transformed again to the necessary service voltage. The current then passes through the meter, flows through the consumer's internal wiring, through the consuming devices, and back through the system to the generator, completing the circuit.

In any electrical circuit, electrons always make their way back to the source generator. Thus, the electrons themselves are not used up. The reason that electrical systems require a continual input of energy (natural gas, coal, water flow, etc.) is simply to keep the electrons moving.

The Key Physical Properties of an Electric Delivery System

There are a number of physical properties unique to electricity that are extremely important to understand. As you will learn later in this book, electricity is a commodity like no other. Understanding these properties will help you to understand why the electricity business operates as it does.

Electricity Cannot be Stored

Flowing electrons cannot be easily stored. This means the electrical system must be operated to ensure that supply and demand are continually in balance throughout the system at all times. If electrical supply is not available to match instantaneous electrical demand, the whole system will crash, resulting in blackouts. Thus an electrical system requires continual surveillance and adjustment to ensure supply always matches demand.

The Path of Electrical Flow is Difficult to Control

Electrons flow on the path of least resistance. And if the least resistant path is from the transmission line through a wet tree branch to the ground, that's precisely where the electricity will flow. Similarly, if this is from one utility's transmission system into another interconnected utility's transmission system, that is where electrons will travel. Thus all utilities on an interconnected system must cooperate in operating their systems as the action of one may cause electrons to flow into or out of the others' systems.

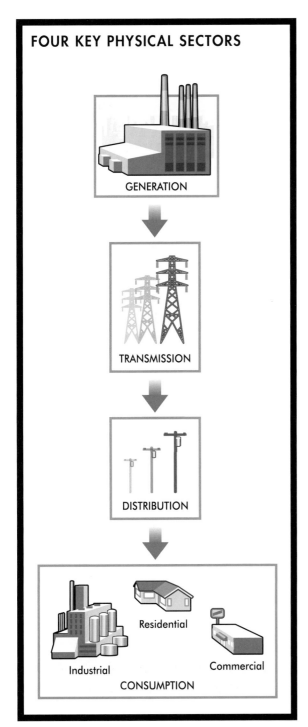

FOUR KEY PHYSICAL SECTORS

GENERATION

TRANSMISSION

DISTRIBUTION

CONSUMPTION
Industrial, Residential, Commercial

Disturbances Travel Very Quickly and are Hard to Contain

Changes in voltage or frequency on electrical lines move at the speed of light, which is 984 million feet (300 million meters) per second. So any disturbance – say a sudden burst of high voltage or a frequency that is out of whack – also travels very quickly. This is why a tree hitting a power line in Oregon can quickly result (and has) in lights going out in Los Angeles. This means that not only must system operators cooperate, but they must also be prepared to react to one another's actions (and problems) almost instantly.

Outages and Significant Voltage or Frequency Fluctuations are Not Acceptable

With the advent of modern electric controls and microchip-based devices, our consumption of electricity no longer tolerates momentary outages or fluctuations in voltage or frequency. Thus the entity that is responsible for matching supply and demand not only has to do so every minute of every day, but must also do it with little margin for error.

There you have it – the four physical properties of electricity that make it different from any other form of energy. These properties not only necessitate a centralized coordinating

SECTION TWO: WHAT IS ELECTRICITY?

function called system operations, they also create business complexities unlike any other business in the world today.

The Four Key Physical Sectors of the Electricity Business

Now that we have studied how electricity is created and delivered, and discussed its key physical properties, we are ready to focus on the four key physical sectors of the electricity business. These are generation, transmission, distribution, and consumption. These are also the four components that comprise the circuit which we call the electric delivery system. Because generation, transmission and distribution are designed to serve specific customer needs, we will discuss electric consumers first. This will be followed by a look at the generation, transmission and distribution sectors. We will explore the business entities and market structures that provide these functions throughout much of the rest of this book.

UNDERSTANDING TODAY'S ELECTRICITY BUSINESS

What you will learn:

- The customer segments that use electricity

- How much electricity each customer segment uses and how use is expected to grow

- How each customer segment uses electricity

- Costs for electricity

- Unique needs by customer segment

- Daily and yearly aggregate usage patterns

SECTION THREE: ELECTRIC CONSUMERS

All business starts with the customer, and the electricity business is no exception. So before we go any further, let's take a look at the various types of electricity consumers (also called end users or ratepayers in the utility industry). Traditionally, end users have been categorized according to the rate classes they were placed in by their local electric utility. These include residential, commercial and industrial classes. Some utilities also have rate classes for agricultural, wholesale and street lighting customers. Wholesale rates apply to customers who are buying power for resale to another customer, but are not an ultimate consumer of the power themselves. These various groups of customers are placed into specific customer classes because they are considered to be "similarly situated" (i.e. their uses for and consumption of electricity are considered to be generally the same). The philosophy of regulation holds that similarly situated customers should pay similar rates.

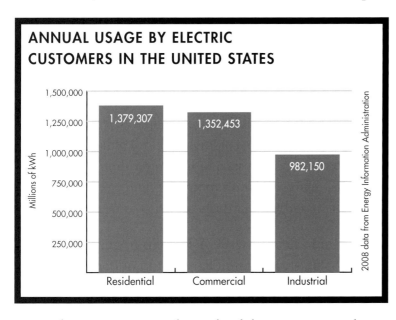

The chart to the left illustrates annual kWh consumption for each of the three major customer classes. You will notice that consumption for each of these customer classes is relatively similar. The number of customers within each class, however, varies dramatically as you can see from the chart on page 22. The number of residential customers far exceeds the number of commercial customers, which in turn far exceeds the number of industrial customers. This will be very important as we consider the rates that these customers pay. A bit of simple math clearly indicates that each industrial customer uses far more electricity than individual commercial or residential customers. This distinction is the primary reason

SECTION THREE: ELECTRIC CONSUMERS

why industrial customers can be served more cheaply, have more clout in the regulatory/political arena, and are generally more attractive customer targets for deregulated electric sales in states where such sales are allowed.

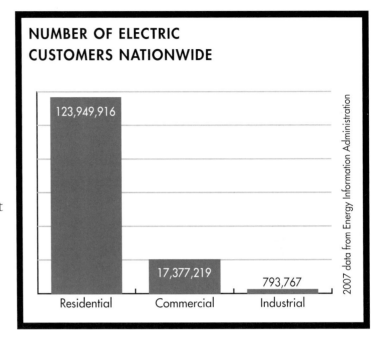

In looking at customer data it is important to remember that market characteristics vary regionally. For example, in the State of Florida residential customers use over six times as much electricity as industrial customers, while nationwide residential customers use only 1.4 times as much as industrial customers. As we will see later, the difference in usage patterns between the three customer classes creates significant differences in how the electrical system must be designed and operated to serve these customers.

Residential Customers

Nearly 124 million residential customers (including single-family homes and multi-family units) in the United States use electricity[1]. As we have seen, residential electricity use accounts for approximately 37% of overall U.S. usage. U.S. residential electricity consumption has increased by about 2% per year over the last ten years. Consumption is expected to continue to increase at about 1% per year over the next 20 years, driven by demand for electricity to power computers and other electronics as well greater use of home air conditioning. Improvements in the energy efficiency of building insulation and lighting have kept increases in usage from being more significant and ongoing energy efficiency efforts may continue to hold down residential load growth.

The top residential uses for electricity include lighting, space cooling, space heating, refrigeration, water heating, TVs, and clothes drying. Other devices that use significant quantities of electricity include freezers, cooking, personal computers, and furnace fans. Because of the significance of space cooling, residential usage is greatest in the summer and early fall months in most regions. During the spring and late fall,

---

[1] Usage and price data in this section is taken from the EIA unless otherwise noted.

usage typically falls by as much as one-third. Winter usage is higher again due to increased space heating and lighting demands. Usage in the coldest months of December and January is frequently as much as 80 to 90% of the peak summer usage.

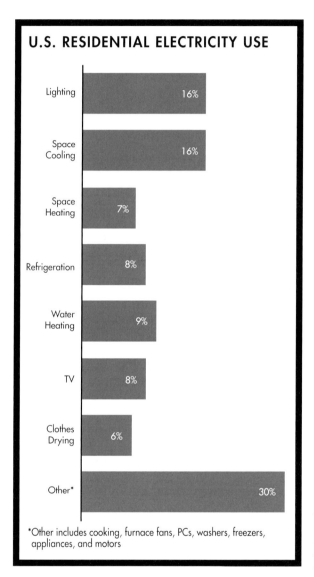

Residential usage also changes significantly from hour-to-hour during the day. Usage is minimal during the late night and early morning hours, rises as residents get out of bed, falls again as residents go to school and work, and then climbs steadily from 3 p.m. through the evening. Often, peak usage late in a day will be more than double that in the middle of the night. Unlike residential consumption of other energy forms such as natural gas, electricity usage from year-to-year does not swing widely. In fact, over the last ten years, U.S. residential energy usage has never varied from year-to-year by more than 5%.

Residential consumption patterns are rarely driven by price in the United States. In most areas, consumers are insulated from wholesale price swings by the nature of electric ratemaking which often averages rates for periods of at least one year. This lack of demand response (to the real-time cost of electricity) has been a contributing factor to wholesale price spikes in competitive markets. In areas where residential consumers are exposed to market prices, demand response has been significantly greater.

Key residential customer needs include:

- Reliability — Residential customers in the United States have come to expect that power will always be available when they need it. Since our society considers electricity a vital resource, even momentary outages are considered a significant nuisance and are generally unacceptable.

SECTION THREE: ELECTRIC CONSUMERS

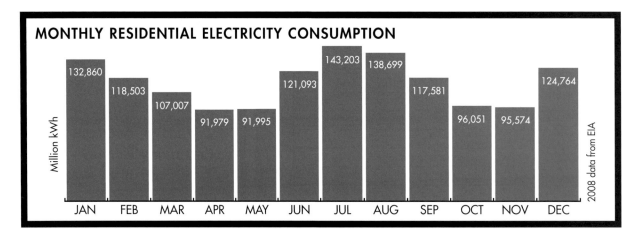

- Power quality to match needs of appliances — Not only do customers need high reliability (no outages), they also need electricity that fits within the voltage and frequency tolerances required by computers and the ever-increasing number of home appliances controlled by microchips. Not too many years ago, low voltages simply meant lights were dim and electric clocks ran slow. In today's world, low voltage means a number of devices will no longer work properly.

- Low prices — The cost of electricity is often a significant factor in consumers' budgets, especially among elderly and low-income customers who commonly have the least efficient homes and appliances.

- Stable prices — The reality of the electricity marketplace is that wholesale prices change hourly (and often within the hour). Almost all residential customers will say that they cannot deal with continually changing price levels. Even monthly price changes have proven to be highly unpopular in areas where introduction of regulatory reforms has resulted in monthly market price adjustments. Most residential customers prefer to lock-in a specific price level for at least a year at a time.

- Energy efficiency — Many residential customers can significantly reduce electric consumption and costs through relatively simple energy efficiency measures. But most don't, due to lack of knowledge or time. Services that make energy effi-

> **LOAD FACTOR**
>
> Load factor is a measure of how average usage relates to peak usage. Load factor is calculated as the total usage for a given period of time divided by what usage would have been had peak demand been maintained throughout the entire period. The following calculation measures load factor over the period of one year:
>
> $$\text{Load factor} = \frac{\text{Actual annual usage (kWh)}}{\text{Peak demand (kW)} \times 8760 \text{ hours}}$$
>
> Load factor is important since electrical services and generation must be designed to serve peak usage levels. During the hours when usage is less than peak, a portion of the facilities installed to meet peak loads sits idle.

ciency improvements more accessible have become important to many customers.

Residential customers typically pay more for electricity than other customer groups. There are three primary reasons for this. First, the distribution system required to serve residential customers is more expensive because services are delivered in small quantities and at lower voltages. Second, residential customers tend to have low load factors. Thus costs for residential customers are generally spread over fewer kilowatt-hours resulting in a greater per kWh cost. And third, customer service costs are often much higher for residential customers since the utility is serving a large number of accounts, and residential customers are often less knowledgeable about their accounts than commercial or industrial customers. Residential customers in the U.S. paid an average rate of $0.1205/kWh (compared to $0.1060/kWh for commercial customers and $0.0717/kWh for industrial customers) in August 2009. It is important to note that rates vary significantly from state-to-state based on regulatory policies, the nature of the required transmission/distribution system and the available resource base.

Commercial Customers

Over 17 million commercial customers use electricity in the U.S. Typical commercial customers include retail establishments, restaurants, motels and hotels, healthcare facilities, office buildings, and government agencies. Commercial customers are often distinguished from industrial customers by size (a typical break point would be a peak demand of 500 or 1000 kW, depending on the utility). Commercial electricity use accounts for approximately 36% of overall U.S. usage and has increased by about 3.5% per year over the last ten years. This increase in consumption is expected to continue at about 1.6% per year over the next 20 years, driven by significant demand increases for computers and office equipment.

AVERAGE RESIDENTIAL COSTS/kWh BY STATE

State	cents/kWh
Hawaii	25.13
Connecticut	20.38
New York	19.12
New Jersey	17.50
Alaska	16.92
California	16.16
New Hampshire	16.09
Massachusetts	15.98
Maryland	15.70
Vermont	15.25
Maine	15.19
District of Columbia	14.45
Delaware	14.37
Rhode Island	13.85
Nevada	13.20
Michigan	12.68
Pennsylvania	12.44
Texas	12.43
Florida	12.26
Wisconsin	12.16
Ohio	11.54
Arizona	11.29
Iowa	11.18
Illinois	11.16
Virginia	11.03
New Mexico	11.01
Georgia	10.91
Alabama	10.81
Minnesota	10.66
Colorado	10.45
Kansas	10.40
South Carolina	10.30
North Carolina	10.29
Mississippi	10.06
Nebraska	9.95
Arkansas	9.87
Missouri	9.74
Indiana	9.59
Utah	9.27
South Dakota	9.18
Montana	9.17
Wyoming	9.15
Tennessee	9.08
Oregon	9.06
North Dakota	8.88
Oklahoma	8.59
Kentucky	8.57
Idaho	8.37
Louisiana	8.12
Washington	8.00
West Virginia	7.92

August 2009 data from Energy Information Administration

SECTION THREE: ELECTRIC CONSUMERS

The top commercial uses for electricity include lighting, space cooling, office equipment, computers, and refrigeration. Other significant uses include ventilation, space heating, water heating, and cooking. Because most commercial usage comes from businesses that run year round, it does not fluctuate nearly as much on a monthly basis as residential usage In fact, usage during the lightest months (late winter and spring) is generally only 20% lower than peak months. Usage within the day also tends to fluctuate less than residential customers since many commercial facilities are used throughout the day and into the evening. Commercial electricity usage from year-to-year is more variable than residential and in the last ten years has varied by as much as 8.5%.

Commercial customers are more likely to be price responsive than residential customers. In fact, many larger commercial customers are well suited to alter demand when prices rise. Through use of various demand shifting or energy efficiency measures, these businesses can reduce electricity consumption when it is cheaper to implement demand management measures than it is to buy electricity. And since it's simply an economic decision for most businesses, commercial customers tend to respond rapidly when proper incentives are in place.

U.S. COMMERCIAL ELECTRICITY USE

- Lighting: 23%
- Space Cooling: 11%
- Office Equipment (non-pc): 5%
- Office Equipment (pc): 5%
- Refrigeration: 9%
- Ventilation: 11%
- Space Heating: 4%
- Water Heating: 2%
- Other*: 30%

*Other includes cooking, telecom, medical equipment, ATMs, and service station equipment

Key commercial electric needs include:

- Reliability — Commercial customers in the United States have come to expect that power will always be available when they need it. Virtually all commercial customers in the U.S. are unable to function without power. Even stores with natural lighting generally cannot operate their cash registers – and thus cannot serve customers – in the event of a power outage. Commercial customers with critical

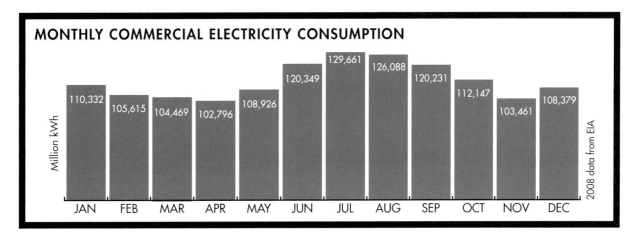

loads such as emergency lighting, food refrigeration, medical equipment, elevators, and telecom equipment often install their own back-up generators or batteries to maintain power should the utility service fail.

- Power quality to match needs of equipment — Not only do commercial customers require high reliability (no outages), they also place a high value on electricity that fits within the voltage and frequency tolerances applicable to computers and the ever-increasing loads of equipment controlled by microchips. In today's world, voltages or frequency fluctuations outside of tolerances can destroy sensitive equipment and bring businesses to a standstill. So even with high quality power service, critical equipment such as computers, cash registers and telecom is still often protected by power conditioning devices that ensure uniform voltages and frequencies.

- Low prices — The cost of electricity is often a significant factor in commercial customers' cost of doing business. Thus they are often very attracted to ways to reduce their electricity costs.

- Stable prices and/or tools to manage electricity costs — The reality of the electricity marketplace is that wholesale prices change hourly (and often within the hour). Most commercial customers cannot effectively deal with continually changing price levels as they are more focused on running their businesses. The exception is larger commercial customers who can manage price exposure through technology. These more sophisticated customers are able to install demand management systems (for example, a smart building management system) that can automatically respond to price signals from the electricity provider.

- Energy efficiency — Many commercial customers pay close attention to energy usage and are willing to invest in energy efficiency measures if they are likely to result in rapid (one year or less) paybacks on investment.

SECTION THREE: ELECTRIC CONSUMERS

Commercial customers typically pay considerably more for electricity than industrial customers, but slightly less than residential customers. Commercial customers are frequently served at slightly higher voltages than residential customers (often 480 V instead of 120 V) and thus require less distribution facilities which reduces distribution costs. They also tend to have higher load factors than residential customers, meaning that the ratio of peak use to average use is smaller, and peak-related costs can be spread over more kilowatt-hours. But customer service costs are still relatively high for commercial customers since the utility serves a large number of accounts, knowledge levels vary and the customer group is highly diverse. Commercial customers in the U.S. paid an average rate of $0.1060/kWh (compared to $0.1205/kWh for residential customers and $0.0717/kWh for industrial customers) in August 2009. It is important to note that rates vary significantly from state-to-state based on regulatory policies, the nature of the required transmission/distribution system and the available resource base.

Industrial Customers

Over 790,000 industrial customers use electricity in the U.S. But don't let this small number fool you – the importance of this sector belies its small number of accounts. While comprising less than one-half of one percent of customers in the U.S., industrial customers consume over 26% of U.S. electrical production.

Typical industrial customers include manufacturing, construction, mining, agriculture, fishing, forestry, electronics, and food processing. Large industrial users of electricity include aluminum, chemicals, cement, forest products, glass, metal casting, petroleum refining, pulp and paper, and steel.

AVERAGE COMMERCIAL COSTS/kWh BY STATE

State	cents/kWh
Hawaii	22.88
Massachusetts	18.51
Connecticut	16.64
New York	16.40
California	15.75
New Jersey	14.91
Alaska	14.25
New Hampshire	14.01
District of Columbia	13.67
Vermont	12.91
Rhode Island	12.72
Maine	12.36
Delaware	12.01
Maryland	11.95
Florida	10.62
Nevada	10.62
Alabama	10.10
Arizona	10.06
Michigan	10.04
Ohio	9.81
Pennsylvania	9.71
Wisconsin	9.70
Texas	9.66
Tennessee	9.28
Mississippi	9.18
Georgia	8.97
New Mexico	8.90
Colorado	8.81
South Carolina	8.78
Iowa	8.72
Minnesota	8.64
Kansas	8.53
Illinois	8.34
Indiana	8.27
Missouri	8.11
Montana	8.11
North Carolina	8.09
Virginia	7.97
Arkansas	7.96
Kentucky	7.87
Nebraska	7.86
Oklahoma	7.75
Louisiana	7.47
Oregon	7.44
Utah	7.43
Wyoming	7.32
South Dakota	7.24
North Dakota	7.23
Washington	7.01
Idaho	6.96
West Virginia	6.56

August 2009 data from Energy Information Administration

As discussed earlier, some utilities use usage levels to classify industrial customers rather than attempting to evaluate end user types. A typical break point between commercial and industrial would be 500 or 1000 kW of peak demand. Unlike other customer groups, consumption by this group has remained steady over the last ten

UNDERSTANDING TODAY'S ELECTRICITY BUSINESS

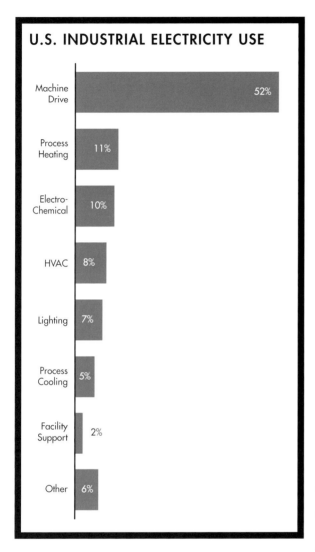

years. Much of the lack of growth was due to large annual decreases in consumption during the U.S. economic slowdowns in 2001 and 2008/2009. Consumption is expected to remain level over the next 20 years. Any demand increase due to greater use of computers, computer-based controls or new technologies is expected to be offset by efficiency gains and movement to less energy intensive industrial output.

Industrial uses for electricity are dominated by machine drives, with other significant uses including process heating, electro-chemical processes, HVAC (heating, ventilating and air conditioning), lighting, and process cooling. Because much of industrial usage is for manufacturing that runs year round, usage does not fluctuate considerably from month-to-month. In fact, usage during the lightest months (mid-winter) is generally only 10% lower than peak months. Usage across the day does not vary as widely as other customer groups since many industrial facilities operate around the clock. Industrial electricity usage from year-to-year swings much more widely than the other sectors due to business cycles. Over the last ten years, U.S. industrial energy usage has remained flat, though with occasional sharp declines by as much as 6.4% in some years, and gains of no more than 2.2% in others.

Industrial customers are more likely to be price responsive than any other customer group as electric consumption is often a significant factor in their cost of doing business. These customers are also much more likely to have alternatives than other customer groups. If electricity rates get too high, manufacturing might be moved to other lower-cost states or even other countries. And some customers have the option to generate their own electricity when it's cost-effective. Many industrial customers are also well suited to price responsive actions with their demands. Demand can often be shifted to lower price periods, energy efficiency measures can significantly reduce electrici-

SECTION THREE: ELECTRIC CONSUMERS

MONTHLY INDUSTRIAL ELECTRICITY CONSUMPTION

Month	Million kWh
JAN	81,331
FEB	79,428
MAR	81,372
APR	81,711
MAY	85,817
JUN	84,855
JUL	85,846
AUG	85,535
SEP	83,200
OCT	82,117
NOV	77,472
DEC	73,464

2008 data from EIA

ty consumption, and many industrial customers can handle periodic interruption of power allowing them to take advantage of cheaper interruptible rate schedules. Industrial customers commonly have the internal expertise and capital to evaluate and implement cost-saving options and/or are able to afford outside energy expertise.

Key industrial customer electric needs include:

- Reliability — Unexpected loss of power can cause costly problems for industrial customers. Because large manufacturing facilities are expensive to build and maintain, industrial customers lose profits whenever their facilities are idle. Many industrial customers back-up critical loads with back-up generators or uninterruptible power supply (UPS) systems. In some cases, industrial customers use their generators to supply some or all of their power needs on site. This insulates them from concerns about utility power reliability.

- Power quality to match needs of equipment — Virtually all industrial processes are now run by electronic control systems. These systems tend to be highly sensitive and can be shut down or even damaged by voltage and frequency fluctuations or spikes. And as we've already seen, unexpected shutdowns can be costly.

- Low prices — The cost of electricity is often a significant factor in industrial customers' cost of doing business. Thus industrial customers have strong incentives to continually find ways to reduce their electricity costs.

- Stable prices — Industrial customers are much more likely to have their electricity prices tied to wholesale market conditions. This is because their large usage more closely resembles a wholesale customer than a retail customer. Some customers can handle the wholesale price fluctuations, others cannot. Large industrial customers are much more likely to take advantage of financial instruments to manage price risk if they cannot get price certainty from their utility or marketer.

- Tools to manage energy costs – Many industrial customers pay prices for electricity that vary from hour-to-hour. Rates may be significantly higher (as much as 100% higher) during peak hours than off-peak hours. Thus industrial customers need real-time information on energy usage and control systems or other technology that allows them to manage their overall energy use during higher-priced periods.

- Energy efficiency — Since industrial customers are huge consumers of electricity (an annual bill of $1 million is not uncommon), small percentage savings in electricity can add up to significant dollars. So industrial customers are continually looking for new ways to boost the efficiency of their electricity use.

- Timely and accurate billing — Because most industrial customers now use sophisticated cost management systems, timely and accurate billing information of their electricity use is very important to their ability to manage their businesses.

Industrial customers typically pay significantly less for electricity than other consumers. Industrial customers are frequently served at high voltages (anywhere from 2.4 kV to as high as 60 kV) and thus require less distribution facilities which reduces distribution costs. Industrial customers also tend to have high load factors, so peak-related costs can be spread over more kilowatt-hours. Customer service costs are relatively low since there are fewer customers and they are more likely to be knowledgeable about their electricity use. Industrial customers in the U.S. paid an average rate of $0.0717/kWh (compared to $0.1205/kWh for residential customers and $0.1060/kWh for commercial customers) in August 2009. Again, it is important to note that rates vary significantly from state-to-state based on regulatory policies, the nature of the required transmission/distribution system and the available resource base.

AVERAGE INDUSTRIAL COSTS/kWH BY STATE

State	cents/kWh
Hawaii	19.13
Connecticut	17.13
Alaska	12.84
New Hampshire	12.77
New Jersey	12.69
Rhode Island	12.34
District of Columbia	11.83
California	11.75
New York	11.53
Massachusetts	10.90
Maryland	9.98
Maine	9.91
Nevada	9.78
Vermont	9.34
Florida	9.29
Delaware	9.13
Michigan	7.91
Illinois	7.46
Arizona	7.29
Pennsylvania	7.13
Ohio	7.10
Colorado	6.89
Wisconsin	6.84
Virginia	6.84
Minnesota	6.83
Georgia	6.61
Texas	6.60
Mississippi	6.50
Tennessee	6.45
Alabama	6.43
Kansas	6.28
North Carolina	6.26
Missouri	6.24
Arkansas	6.23
Iowa	6.12
Nebraska	6.01
South Carolina	5.96
North Dakota	5.93
Indiana	5.80
Oklahoma	5.75
Idaho	5.74
New Mexico	5.74
Montana	5.53
South Dakota	5.52
West Virginia	5.41
Oregon	5.40
Kentucky	5.34
Washington	5.20
Utah	5.19
Wyoming	5.16
Louisiana	4.81

August 2009 data from Energy Information Administration

SECTION THREE: ELECTRIC CONSUMERS

Aggregate Demand Curves

Because electricity cannot be stored, electric providers must be prepared to service the total demand from their customers at all moments during the day. This is true for utilities in regulated states that must plan to have enough generation available to serve their customers, it is true for Independent System Operators (ISOs) that must plan to have enough generation available to serve the entire marketplace, and it is true for retail marketers that must plan enough supply to match their contracted customers' demand as closely as possible or risk significant financial exposure for balancing costs. Thus it is important to understand aggregate customer behaviors. Unlike the gas business, where transactions tend to take place on a daily basis, electricity sales and price fluctuations often occur in increments of one hour, and for balancing power, in increments as short as five minutes. So hourly and even intra-hourly usage fluctuations are important to clearly understand.

A typical hourly load curve might look like this:

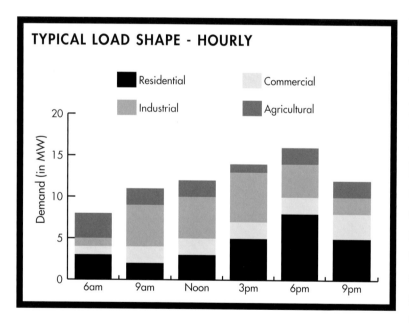

Observe the change across the day for the various customer classes. The agricultural sector peaks in the early morning because farmers do their water pumping before the heat of the day. Residential usage climbs in the morning as residents arise, falls as they go to work and school, and then climbs through the afternoon and early evening as they return home. Commercial and industrial uses are steadier during the day, but fall significantly at night as many businesses close.

Seasonal patterns are also important since some customers' electricity consumption varies greatly by season. The annual seasonal load curve for the U.S. looks like this:

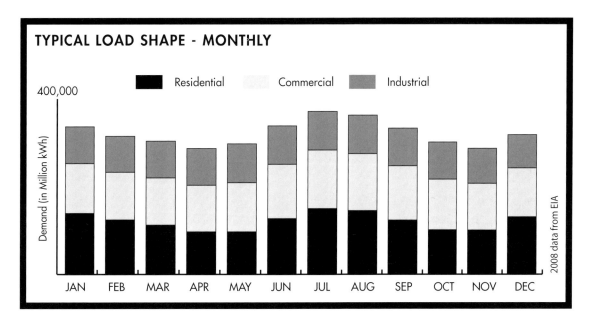

As mentioned earlier, customer mixes and weather patterns vary significantly regionally. And since electric markets are constrained to regional areas by the lack of transmission, it is critical that industry players understand the consumption patterns of each region in which they are active.

What you will learn:

- What generation is
- The different types of electric generation
- The characteristics of each generation type
- The costs associated with each generation type
- Environmental concerns with each generation type
- How demand response serves as an alternative to generation
- How different generation sources are used to meet the demand curve
- Who owns generation
- How utilities and generation companies evaluate needs and develop generation portfolios
- Future generation sources

SECTION FOUR: GENERATION

Now that you have a general understanding of the needs of various electric customers, it's time to turn our attention to the physical system that was designed to deliver service to them. We will begin with a discussion of generation, which is the creation of flowing electrons. In later sections we will consider the other key components of the delivery system – transmission, distribution and system operations.

Generation fuel sources in the United States are quite diverse and include coal, nuclear, natural gas, fuel oil, hydro, and various forms of renewable energy. Within any of these fuel sources, the technology employed to generate electricity can be diverse as well (for instance, gas technology includes steam turbines, combined-cycle turbines and single-cycle combustion turbines). Each type of generation has unique operating and cost characteristics that make it more or less suitable to a specific supply need. This is why utilities or generating companies generally build generation portfolios comprised of varied generation types. These can then be used to match the needs of their customers as well as the specific needs of their geographic location.

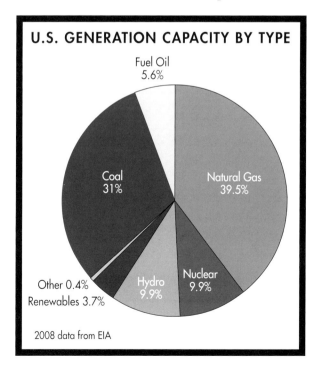

U.S. summer generation capacity is approximately 1,008,800 MW. The United States' generation portfolio that provides this capacity is comprised of many types of power plants with natural gas plants being the most common. But because coal and nuclear units are typically used for what is called baseload, they are run more frequently than other types of generation. Thus the generation output percentages look significantly different as you can see in the chart on page 36. Note that coal output rises to the top with 49% of U.S. output in 2008, followed by natural gas at 21%

SECTION FOUR: GENERATION

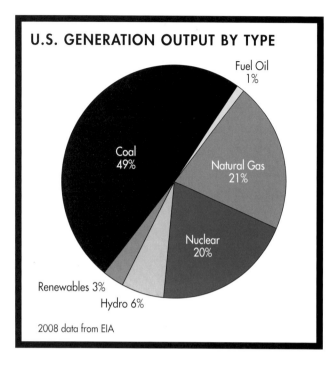

and nuclear at 20%. These percentages vary from year-to-year based on factors such as the cost of natural gas and fuel oil, the amount of hydro power available due to weather conditions, and the total amount of consumption.

U.S. generation capacity grew by an annual average of 2% per year for the five years from 2002 to 2007. The majority of this growth was due to construction of natural gas-fired units with additional significant growth added by renewable energy. Late in the 2000s utilities and merchant generators began a new wave of generation construction resulting in additional coal, gas and renewable projects, perhaps followed by new nuclear capacity by 2020.

Types of Generation

As you know, utilities or generating companies try to match generation types with the aggregate needs of their customers. To understand how this is done, it is important to first understand that each generation type has different operating, financial and environmental characteristics. Key characteristics include capital costs, variable costs, operational flexibility, environmental impacts, fuel availability, and restraints on locations where units can be constructed. Following is a discussion of each generation type and an assessment of the key characteristics outlined above. It is important that you clearly understand these fundamental characteristics as they will dictate which units run when and the generation type and technology used in the construction of new plants.

Coal

The ready availability of low-cost coal has historically made coal-fired generation a favorite of many U.S. utilities. Most coal-fired generation employs steam turbine technology where coal is burned to heat water in boiler tubes. The water becomes steam and is run through a steam turbine that drives a generator shaft to create electricity. Because of economies of scale, most coal units are fairly large – in the range of 250 to

UNDERSTANDING TODAY'S ELECTRICITY BUSINESS

> **GENERATION CHARACTERISTICS**
>
> **Capital cost** — The up-front costs associated with buying equipment and constructing the generation unit, expressed in $/MW capacity.
>
> **Variable cost** — The costs associated with running a generation unit that are directly related to the unit output, including fuel, water and maintenance. These costs are expressed in $/MWh.
>
> **Operational flexibility** — How quickly can a unit be turned on or off, and how quickly can it ramp from low power to full power?
>
> **Time to permit and construct** — How long it takes to permit and build a new unit.
>
> **Environmental impact** — What environmental impacts result from construction and operation, and what is the cost of environmental mitigation?
>
> **Fuel availability** — How certain is future fuel supply?
>
> **Locational** — Is location of unit constrained by fuel or water availability, land use concerns or transmission availability?

1500 MW. The capital costs associated with building coal units are generally high compared with gas units, but many existing units have been on-line for a number of years and thus have been significantly depreciated. Operations and maintenance (O&M) costs are relatively low depending on the age of the unit. Fuel costs have tended to be among the lowest of generation sources in the U.S. Due to technological constraints, coal units have limited operational flexibility. If the unit is running at partial power it can often be ramped up or down in response to system needs. But if the boiler has gone cold, it will require several hours to get to full operation. Because burning coal can be responsible for considerable emissions, coal units are generally considered to have a higher environmental impact than other sources of generation. For this reason and because of high transportation costs, there are areas of the country that do not use much coal to generate electricity. Other areas are highly dependent on coal generation.

Beginning in the mid-2000s a number of utilities and merchant generators planned for construction of new coal units. But rising construction costs, falling natural gas prices, and opposition due to concerns over emissions led to cancellation or postponement of many projects. While ongoing projects will lead to significant new coal capacity over the next few years, a number of older units may be retired, likely resulting in only a small net increase in capacity. Some companies are considering clean coal technologies such as those used in Integrated Gas Combined-Cycle (IGCC) units as a means of tapping our plentiful coal resources while reducing environmental impact. Significant research is also being devoted to developing economic ways to remove greenhouse gases from existing coal units.

Nuclear

A number of nuclear units were brought on-line in the United States in the 1970s and 1980s. These units are generally large and range in size from 600 to over 1200 MW. Nuclear generation uses the heat of nuclear fission to create steam that is then run through a steam turbine. Capital costs associated with new nuclear units are very high, but as the units age and are depreciated their book values have declined. Variable costs including fuel are generally low, but fixed maintenance costs are higher due to the extreme safety procedures required as well as the need to collect costs for future decommissioning. Because of the technology employed, nuclear units do not have good operational flexibility, and start-up times are usually measured in days. Because of this inflexibility, nuclear units are used for baseload needs. New development of nuclear generation in the U.S. has been hampered by two key issues – the lack of a waste disposal site for spent fuel and public concerns over the risks of a major nuclear accident or terrorist attack. In fact, no new units have been brought on-line in the U.S. since 1996 (although new nuclear units have continued to be built in other countries). As of late 2009, a few companies had begun the licensing process for new nuclear units in the U.S., and one project previously suspended had moved back into the initial stages of construction. Despite the perceived safety issues, nuclear generation is favorable from the standpoint of emissions – no greenhouse gasses or pollutants such as NOx, SO_2 or Mercury are emitted from nuclear generation.

Natural Gas

As we have seen, a very high percentage of new generation built in recent years in the U.S. has been natural gas generation. There is also a large base of older gas-fired steam turbine units in the U.S. generation portfolio. Gas-fired generation makes use of three primary technologies – combustion turbines that use natural gas directly to fire a turbine which drives the generator shaft; steam turbines that

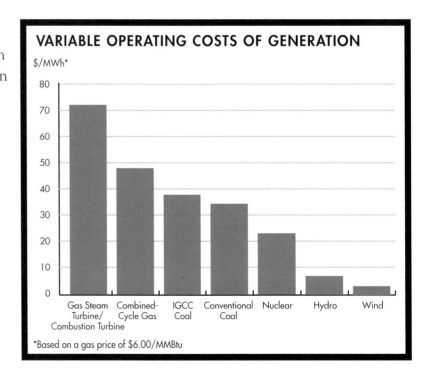

> ## HEAT RATE
>
> The heat rate of a generating unit is a means of measuring the efficiency of the unit by answering the question "How much fuel is required to generate a kWh of electricity?" Other factors being equal, a smaller heat rate is better, since this means less fuel is being consumed to create a unit of electricity. Typical heat rates for various generation types are as follows:
>
> | Unit Type | Typical Heat Rate |
> |---|---|
> | Natural gas steam turbine | 10,000 – 12,000 |
> | Natural gas combined-cycle | 7,000 – 8,000 |
> | Natural gas peaking turbine | 10,000 – 12,000 |
> | Coal steam turbine | 9,000 – 14,000 |
> | Natural gas turbine with cogeneration | 5,000 – 6,500 |
> | Natural gas microturbine | 12,000 – 14,000 |
> | Natural gas fuel cell | 5,000 – 7,500 |
>
> The variable fuel cost of operating a unit can be determined by multiplying the cost of fuel by the heat rate (and usually making some unit conversions). For instance, a natural gas combined cycle-unit with a gas cost of $5.00/MMBtu and a heat rate of 8,000 Btu/kWh will have a fuel cost of $40/MWh.

burn natural gas to create steam in a boiler which is then run through a steam turbine; and combined-cycle units that utilize a combustion turbine (fired by natural gas) and then a steam turbine (wherein waste heat from the combustion turbine is used to produce steam which is then run through the steam turbine). Utility-owned natural gas units vary significantly in size, ranging from as small as 1 MW to over 500 MW. Natural gas is also used to fuel on-site cogeneration units and backup generators for many buildings. Capital costs associated with natural gas units are considerably lower than other generation sources. O&M costs are also generally low. Fuel costs vary depending on the market value of natural gas. As you might imagine, a major concern among owners of natural gas generation are the recent fluctuations in natural gas prices. Depending on technology, natural gas units can be very flexible operationally. Combustion turbines, often called peaking turbines, can be started and stopped within minutes. Steam turbines may require up to six hours to go from cold status to full power. Although gas units do have some air quality impacts, they are generally less harmful than other carbon fuels (coal or fuel oil) and thus considered favorable from an environmental standpoint. For this and other reasons (units can be smaller, easy access to fuel supply, etc.) gas units also have the advantage that they can be located closer to major loads, and thus require less transmission.

Hydro

Hydro power is the backbone of many electric generation systems across the United States where significant hydro resources are available (notably the West and parts of the Southeast). Hydro power is created by running water from a reservoir through a hydraulic turbine that spins and drives a generator shaft. Because the power output can be controlled by simply adjusting the water flow, hydro units are generally very flexible. Hydro units range from very small (100 kW) to very large (over 500 MW) with many units in the 100 MW range. Most hydro units were built a number of years ago (with some units dating back to the 1920s), so capital costs have

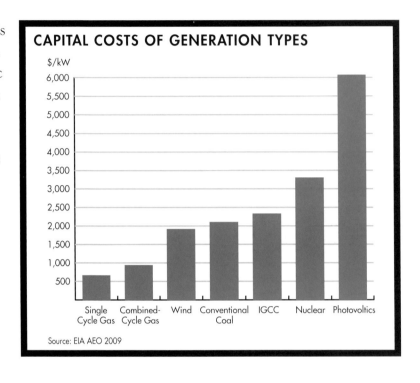

generally been depreciated. O&M costs are generally low and, of course, there is no fuel cost once water rights are acquired. Given their operational flexibility, hydro units are very useful for managing peak loads and for power regulation purposes (keeping supply and demand in balance minute-by-minute) as well as for restoring the grid after a blackout. Although a new hydro dam would now be considered to have large environmental impacts, existing units are generally considered environmentally favorable, with the exception of concerns over impacts on fish populations and downstream activities. A related technology is pumped hydro storage which uses off-peak power to pump water uphill into a reservoir, thus making it available for generation during peak hours. This process is used by utilities as one of the few forms of electricity storage available to them.

Fuel Oil

A limited number of utilities make use of fuel oil generation as an alternative to natural gas. Fuel oil generation is typically seen in regions where natural gas supply is limited or where utilities have the capability of fuel-switching units based on the relative price of fuel oil as compared to natural gas. The technology used in fuel oil generation is similar to natural gas with a few changes to account for physical characteristics of the different fuel. Thus operational characteristics of fuel oil units are similar to natural gas units. The major drawback to fuel oil units is that they have more environmental impacts than their natural gas counterparts. In fact, some areas of the country do

UNDERSTANDING TODAY'S ELECTRICITY BUSINESS

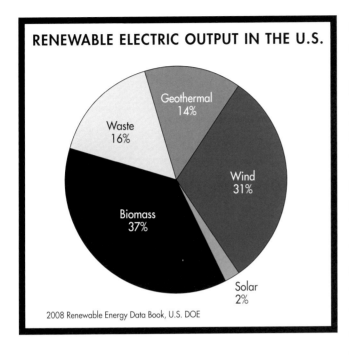

2008 Renewable Energy Data Book, U.S. DOE

not permit fuel oil generation due to air quality concerns.

Renewables

Renewable electricity generation is a broad category fueled by sources that can be naturally replenished. These include geothermal, solar, wind, biomass, and smaller-scale hydro generation (usually less than 30 MW). Technologies used vary widely (see box below on renewables), and the size of renewable units tends to be small. Capital costs per unit vary greatly. Wind power projects are often cost-competitive with natural gas or coal units on a 10-year overall cost basis, while other sources such as solar power may be significantly more expensive. O&M costs vary greatly by technology as well. The two big advantages of renewable power are that many technologies have no ongoing fuel cost (wind, solar) and environmental impacts are generally minimal. One drawback is that many renewable

---

RENEWABLE ENERGY

Renewable electricity generation is fueled by sources that can be naturally replenished. These include:

Biomass — Organic non-fossil fuel that is burned directly to create steam for a steam turbine, or biogas used in a gas turbine. Biogas is created by decomposition of organic material at landfill or agricultural sites and is often at least 50% methane.

Geothermal — Hot water or steam extracted from underground reservoirs in the earth's crust that is used to drive steam turbines.

Hydro — Small scale hydro generation, often run-of-the-river (meaning no reservoir is created), and usually less than 30 MW in size.

Solar — Sunlight applied to a photovoltaic cell (a substance that directly converts light energy to electricity) or sunlight used to heat water and create steam which is then used in a steam turbine.

Wave — Electric generators driven by the movement of ocean water.

Wind — Electric generators whose shaft is driven by the force of wind across a wind turbine.

SECTION FOUR: GENERATION

sources are intermittent and not always available. Thus system operators must plan for backup sources when including renewables in the generation mix.

Development of renewable generation – especially wind power – has accelerated in recent years. Factors fostering this development include concerns over future gas prices, uncertainties over future environmental mitigation costs for coal, a significant reduction in the capital costs of wind power projects, renewable portfolio mandates in some states, and green power programs offered by some utilities and retail marketers (which allow consumers the choice to be served by green power).

Distributed Generation

Distributed generation refers to generation that is located at an end-use consumer's site. Common technologies include generators driven by gas turbines or internal combustion engines. Included in distributed generation are cogeneration units. Cogeneration units, also known as combined heat and power (CHP) generators, are units that utilize their energy input to create two forms of useful energy – electricity and heat. A common application would be to generate electricity using a combustion turbine or internal combustion engine, and to recapture the waste heat from the gen-

GENERATION CHARACTERISTICS COMPARISON							
	Coal	Nuclear	Natural Gas	Hydro	Petroleum	Renewables	Cogeneration
Capital Cost	Medium	High	Low	High	Low	Medium - High	Medium
Variable Cost	Low	Low	High	Low	High	Low	Medium
Operational Flexibility	Medium	Low	Medium - High	High	Medium - High	Low	Low
Time to Permit and Construct	Long	Long	Short	Long	Short	Short	Short
Environmental Impact	High	Low	Low	Low - High	Medium	Low	Low
Fuel Availability	Plentiful	Plentiful	Some concerns	Limited	Some concerns	Plentiful	Some concerns
Availability of Sites	Limited	Limited	Flexible	Limited	Limited by air quality permits	Limited by resource availability	Flexible

erator and use it in direct heating or producing steam needed for internal processes. Alternatively, steam may be created in a boiler, used to drive a steam turbine, and then used for internal process needs. Because cogeneration makes dual use of the fuel, and because distributed generation does not result in transmission or distribution losses, cogeneration can be a highly efficient way to use fuel. Some CHP applications create efficiencies as high as 85% compared to the U.S. electrical average of about 35%. Location at the customer site can also reduce transmission and distribution costs and enhance local reliability. Future distributed generation may use new technologies such as fuel cells and micro CHP (small cogeneration units with a capacity of about 1 kW which is a good baseload for a residence).

ENVIRONMENTAL CONCERNS		
Category	Specific Issue	Environmental Impact
Air Pollution	Sulfur dioxide (SO_2) Nitrogen oxides (NOx) Carbon dioxide (CO_2) Mercury	Acid rain, local health issues Smog Global warming Local health issues
Water Resources	Use of water Thermal discharges River ecosystem disruption	Consumption of water resources Damage to fish and other species Damage to fish and other species
Nuclear Radiation	Release of radiation from fuel Major accident radiation release	Possible source of cancer Source of cancer and other diseases
Land Use	Disrupted environments – mining Disrupted environments – construction	Impacts on pristine areas Visual and economic impacts in urban areas, disruption to pristine land in rural areas

Environmental Considerations

The generation of electricity results in an environmental conundrum – use of electricity at the point of consumption is very clean (for instance, electric cars are non-polluting) yet generation of electricity often has significant environmental impacts. These include air pollution, water pollution, greenhouse gas emissions, ecosystem and land-use disruption, and the potential for release of radioactive materials (see table above). Areas of greatest concern include electric generation's contribution to acid rain, smog, global warming, and local health issues, as well as the potential for radiation release.

ENVIRONMENTAL ISSUES BY GENERATION TYPE

Generation Type	Environmental Issues
Coal	• CO_2 • NO_x • SO_2 • Mercury • Other heavy metal particulates • Land use disruption for mining • Use of diesel locomotives to transport coal
Nuclear	• Low level radiation release through mining and waste • Potential for high level radiation release in an accident
Natural Gas	• NO_x • CO_2 (but much less than coal) • Land use disruption for drilling
Hydro	• Impact on downstream fish and other species
Renewables	• Land use disruption for wind power

Different types of generation have very different impacts, and environmental considerations can greatly influence how generation types are used as well as what types continue to be built. For example, environmental mitigation costs create an unattractive uncertainty for coal generation. Similarly, the potential for future political/environmental issues associated with nuclear generation have prevented any nuclear unit construction in the U.S. in the recent past. To foster cleaner generation sources, some states have moved to renewable portfolio standards (RPS) that require utilities and/or generation providers to acquire a certain percentage of their generation portfolio from renewable resources. Meanwhile, many utilities and generating companies favor new construction of gas-fired units in part due to the relative ease in obtaining environmental permits.

Electric Generation, Global Warming and the Kyoto Protocol

The world's scientific community and most of the world's political community now agree that man's activities in burning carbon-based fuels (coal, petroleum and natural gas) is resulting in raised concentrations of greenhouse gasses which increase the earth's average temperature and destabilize weather patterns. If the trend continues, results could be severe and include melting of ice packs, flooding of low lying areas, interruption of food production, and increased incidents of severe weather.

The largest single greenhouse gas is carbon dioxide (CO_2). About 40% of the CO_2 emitted in the United States (which is responsible for about one-quarter of the world's CO_2) is the result of electric power production. More of this is from coal generation

UNDERSTANDING TODAY'S ELECTRICITY BUSINESS

ENVIRONMENTAL REGULATION OF POWER PLANTS

Federal Regulation

The Clean Air Act lays out many of the rules concerning environmental regulation of power plants. All new generating units must undergo review and meet the requirements of this legislation. Pollutants currently regulated include NOx and SO_2, and the federal government has recently proposed new rules concerning mercury. NOx and SO_2 are regulated under "cap and trade" programs that allow utilities and others to trade federal allowances. This means units with cleaner emissions can trade unused credits to units with worse emissions.

The Clean Air Act requirements do not apply to plants built prior to 1978. Thus many older coal plants may have emission rates that are as high as ten times those of newer units which use advanced control technologies. Regulations developed under the Clean Air Act do not currently regulate CO_2 or other greenhouse gas emissions, but the U.S. Supreme Court recently ruled that the U.S. Environmental Protection Agency has the authority to regulate greenhouse gasses under the Act.

State Regulation

Through their regulation of generation supply planning for vertically-integrated utilities, states have traditionally had significant involvement in evaluating the environmental impacts of generation resources. Prior to permitting and construction, a power plant must undergo a state environmental impact review. Some states are also beginning to implement regulation of greenhouse gasses in the absence of federal standards.

than any other source (coal generation emits about twice as much CO_2 per unit of output than natural gas generation). Control technologies for CO_2 emissions from traditional power plants are under development, but not currently commercially available.

In December 1997 a number of countries agreed to the Kyoto Protocol – a historic agreement that is designed to reduce greenhouse gas emissions by establishing national emissions limits and provide for global trading in greenhouse gas emissions credits. The major industrial powers of the European Union, the United States and Japan agreed to cut emissions by 8%, 7% and 6% respectively below 1990 levels over a five-year period beginning in 2008 and to create a global emissions credit trading program. As with all treaties, the Kyoto Protocol requires ratification by each signatory country. In the United States, the Senate is required to ratify treaties. After the Presidential election in 2000, the Bush Administration announced that it would not support the Kyoto Protocol and never requested that the Senate ratify it. The Kyoto Protocol was ratified by the majority of European countries, Japan and Canada, and went into effect in February 2005. The participating countries are now moving forward with implementation of greenhouse gas caps and international trading of greenhouse emissions credits. The Kyoto Protocol will expire in 2012. As of late 2009, governments representing over 170 countries were involved in talks to negotiate a successor agreement.

SECTION FOUR: GENERATION

Meanwhile in the U.S., a number of states are moving forward with regulation of greenhouse gases, and it appears federal regulation will be implemented at some point in the future, creating significant uncertainty for U.S. power generators. It is unknown whether there will be future costs associated with greenhouse gas emissions, and if so, what the level of these costs will be. And since U.S. generators will – at least initially – be unable to participate in the Kyoto Protocol trading of emissions credits, it will be more difficult for them to hedge the risk of potential future costs.

Demand Response as an Alternative to Generation

An alternative to some generation, especially to peaking units, is to develop mechanisms for end-use demand reduction during peak times. Such programs are often called demand response. Demand response can be emergency demand response (where customers are required to reduce demand only during times when their failure to do so will create reliability issues) or economic demand response (where customers are given economic incentives to reduce demand during times when it is cheaper to reduce demand than to purchase or generate additional units of electric supply).

In the 1980s utilities implemented demand programs, typically called demand side management or DSM, whose goal was to reduce the need for costly new generation construction. These programs encouraged customers to implement energy efficiency measures through rebates for more efficient appliances and offered incentives such as discounted curtailable rate schedules which allow the utility to curtail service during times when high demand threatens system reliability. Although many of these programs still exist, there has been a trend lately towards economic demand response (EDR) programs. EDR recognizes that demand response to high prices can have a significant impact on muting price spikes in competitive markets. Thus, utilities and retail marketers have an interest in creating means by which customers can be compensated for reducing demand during high price times – even when reliability is not a factor. Traditional rates that do not pass real-time price signals to customers fail to incite this behavior. EDR programs include:

- Real-time pricing — Customers pay hourly prices that reflect same-day or day-ahead market conditions.

- Voluntary load response — Customers are offered a payment for curtailing blocks of load, usually in the day-ahead.

- Curtailable capacity call — Customers are paid a capacity payment to give the utility or marketer the right to curtail blocks of load under certain conditions; failure to curtail results in payment of market rates for that block of load.

- Automatic load response — Customers are paid a capacity payment to give the utility or marketer the right to remotely and automatically curtail blocks of load.

It is expected that as the cost of generation and concerns over environmental impacts grow, economic demand response will become an increasingly common option for meeting peak power requirements.

Use of Generation to Satisfy the Load Curve

The key to understanding electric supply markets is to understand which generating units are dispatched at what times to meet the load curve (the aggregate demand of all customers in a specific region). This has a major impact on wholesale electric markets since the cost of the last unit required in any given hour (the marginal cost) often determines the market price of electricity in competitive markets.

Generating units are typically scheduled hourly (one day in advance) based on least-cost supply subject to reliability, operating, locational, and regulatory constraints. Use of generation is often divided into three categories – baseload, which is generation run all twenty-four hours of the day; intermediate, which is run from mid-morning until the evening; and peaking, which is run during the peak hours (often from early afternoon until early evening). Units typically scheduled include:

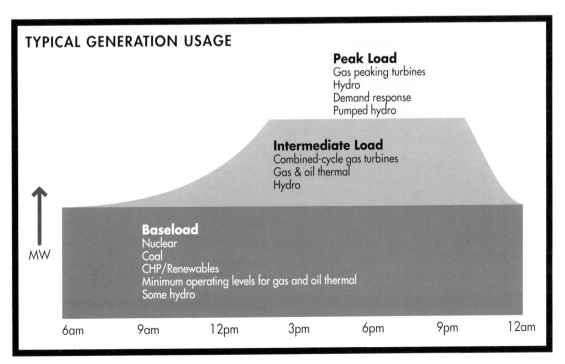

Baseload

Baseload is typically satisfied by nuclear and coal generators (due to their low variable costs and limited operational flexibility), hydro generation (due to low variable costs) and Qualifying Facilities (due to regulatory requirements). In addition, system operators may schedule as baseload gas and oil steam turbine generators that need to be run to provide locational support to the grid (known as must-run generation). It is also sometimes necessary to run gas or oil steam turbine generators at minimum loads because their full capacity will be needed later in the day and their boilers must be kept warm so that the units can be ramped up for availability during the intermediate period.

> **WHAT IS A QUALIFYING FACILITY (QF)?**
>
> In 1978, the U.S. Congress passed the Public Utilities Regulatory Policy Act (PURPA), which contained measures to encourage more efficient use of energy resources. Among these provisions was a requirement that utilities buy the output of qualified cogeneration resources at the host utility's avoided cost rate, which is the cost the utility would pay to generate replacement power if the QF did not exist.
>
> Pursuant to PURPA, the Federal Energy Regulatory Commission (FERC) set forth criteria for determining which facilities could receive Qualifying Facility (QF) status. To be a QF, a generating facility must produce electricity and another form of useful thermal energy (such as heat or steam) used for industrial, commercial, heating, or cooling purposes and must be less than 50 MW in size and meet certain ownership, operating and efficiency criteria. Since investor-owned utilities are regulated by the states, FERC left it to them to define the avoided cost rate. Some states such as California initially set attractive QF rates resulting in significant development of cogeneration facilities (as much as 16% of the California ISO's supply comes from cogenerators), while other states set much lower rates and did not see much in the way of cogeneration development. QFs are typically located at large industrial facilities in industries such as food processing, refineries and wood processing.
>
> The Energy Policy Act of 2005 resulted in changes in rules for new QFs so that in competitive markets they will no longer receive regulatory-set pricing.

Intermediate

Intermediate loads are often satisfied by gas and oil steam turbines, combined-cycle gas turbines and hydro power. These are used because their operational flexibility allows them to be ramped up and down as loads rise and fall during the day, and also because their variable costs are lower than other options.

Peaking

Peaking loads are usually satisfied by single-cycle gas turbines (also known as peaking turbines), hydro power, pumped hydro where available, and economic demand response.

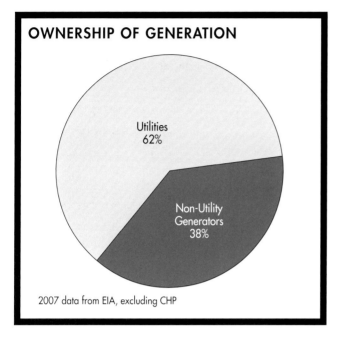

OWNERSHIP OF GENERATION

Utilities 62%
Non-Utility Generators 38%

2007 data from EIA, excluding CHP

As we will learn in a later section, system operators must also schedule generation reserves to ensure system reliability. Good sources for reserves include hydro power, gas turbines, and steam turbine units that are running at partial capacity.

Ownership of Generation

Prior to electric restructuring, virtually all the generation in the United States was owned by either investor-owned utilities, public utilities and rural co-ops, or by federal agencies. Since the advent of electric deregulation, a new category of generation owner has entered the market. These are non-utility generators, which include independent power producers (IPPs) and merchant generators. Non-utility generators are companies that own generation as a stand-alone business and not as part of a vertically-integrated utility. These generators aim to own and operate their units for a profit by selling energy or capacity to utilities, marketing companies and/or directly to end-use customers. In some states, utilities

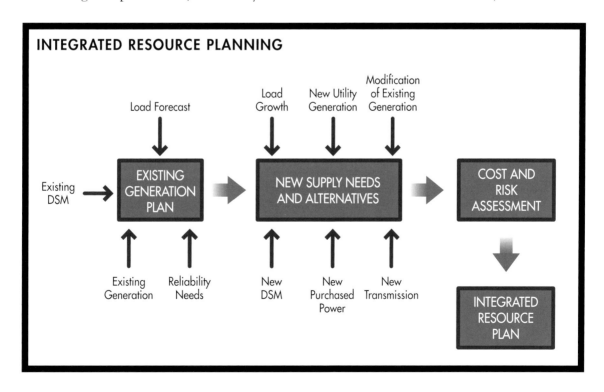

INTEGRATED RESOURCE PLANNING

have divested of their generation assets by either selling them to non-utility generators or splitting the assets off from their utility function and creating their own non-utility generator subsidiary. In other states, non-utility generator ownership is limited to new units that have been built in recent years.

Developing a Generation Portfolio

Companies that own generation must determine the best portfolio of generation units to satisfy the energy and capacity needs of their customers. In the past, this was a relatively simple task. Utilities historically started with their existing generation base, compared this against load forecasts and then evaluated options to fulfill any additional supply requirements. Options might include modifications to existing utility generation, new utility generation, demand side management programs, purchased power (from neighboring utilities or non-utility generators), and new transmission lines to areas with excess generation. These options were then evaluated from the standpoint of cost and risk resulting in an integrated resource plan (see illustration on page 49). Integrated resource planning is used by utilities that have supply responsibilities to evaluate options for obtaining additional supply. Integrated resource plans are filed with the state utility commission for approval before being implemented.

Non-utility generators must also evaluate generation portfolios, but their evaluation focuses on how to best create a return on shareholder investment. Thus a non-utility generator would carefully evaluate the market value of specific generating units as well as the strategic value of units relative to the company's other assets and business strat-

FACTORS AFFECTING THE VALUE OF GENERATION

- Physical flexibility of unit
- Expected O&M costs
- Fuel efficiency
- Future price expectations for fuel
- Future price expectations for energy
- Future price expectations for ancillary services
- Expected price volatility and price spikes in each market
- Environmental and operating permit risk/benefit
- Location relative to transmission capacity
- Opportunity for reliability must run contracts
- Opportunity to expand at site
- Impact on your company's sales strategies
- Impact on your company's risk portfolio
- The applicable discount rate of capital

egy (see box on page 50). Rather than an integrated resource planning process, non-utility generators use portfolio theory to attempt to maximize returns relative to perceived market risks.

In either case, generation planners must evaluate future load growth and the capability of existing generation to satisfy that growth, and then look for points in the load curve where new generation is needed.

The Future of Generation

In the early 2000s, the U.S. experienced a period of significant electric generation construction. 27,000 MW of new capacity was brought on-line in 2000 (the first time that more than 20,000 MW of new capacity had been brought on-line in a single year since 1985), followed by 42,000 MW in 2001, 72,000 MW in 2002, and 43,000 MW in 2003. Generation construction has since slowed and it is expected that additions of generation capacity at these levels will not again be seen for many years.

As we saw earlier in this section, most of the recent capacity additions have been gas units. In fact, 83% of capacity additions from 2002 to 2007 were gas-fired. Most of the remaining new generation has been renewables led by wind power. Although still small in absolute numbers, the amount of wind generating capacity in the U.S. has more than tripled in this same time period.

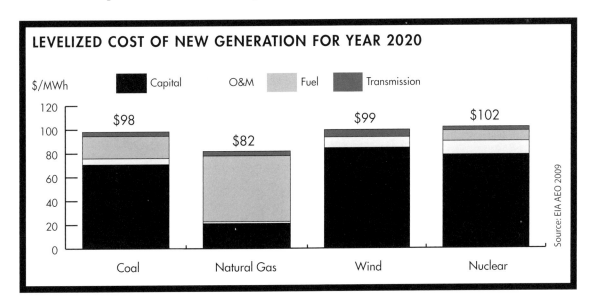

SECTION FOUR: GENERATION

Given the lead times necessary to construct new generation, utilities and some non-utility generators are already planning for the next wave of power plant construction. As you have learned, a key question is what type of units to build. Considerations include capital costs, ongoing O&M and fuel costs, and risks associated with fuel cost volatility and future environmental mitigation. The recent decrease in natural gas prices (natural gas prices at the wellhead fell from over $10/MMBtu in summer 2008 to less than $3/MMBtu in fall 2009) coupled with reports of new sources of new U.S. natural gas supply sufficient to satisfy expected demands for many years into the future has made many participants refocus on construction of natural gas units. While a number of new coal units have recently or will soon be completed in the U.S., many others that were planned have been cancelled or postponed. The future of new coal generation is uncertain. Given robust coal resources it appears likely that more coal generation will be built in the U.S. But technological changes to reduce emissions will likely be required before significant capacity is built. Interest in nuclear generation has grown in recent years due to its lack of emissions and applicability as a baseload resource. A number of companies have applied for new permits to construct nuclear units, but how many will actually proceed to construction is uncertain. Growth in renewable energy is expected to continue. Much of this will be wind power, but in some areas geothermal and solar growth is expected to be significant.

NAMEPLATE GENERATION CAPACITY ADDITIONS (PLANNED FOR 2008-2012)

Type	MW
Coal	23,347
Fuel Oil	1,190
Natural Gas	48,100
Renewables	18,110

Source: EIA

The possibility for technological innovation in generation continues. The last big innovation – the use of combined-cycle gas technology – resulted in a new wave of highly efficient and relatively clean generation across the country. New technologies that may have major impacts include integrated gas combined-cycle units (IGCC) and fuel cells.

IGCC units use coal as a fuel, but eliminate many of the environmental concerns by gasifying the coal prior to combustion. This improves the efficiency of the combustion process and results in emissions that are similar to natural gas units. It is also possible through additional processes to remove most of the carbon prior to combustion, making IGCC even more environmentally attractive. The carbon can then be sequestered by storing it underground to prevent its release into the atmosphere. Many in the utility industry are optimistic about IGCC (a few test units are currently running) and believe this will become an important technology in the next ten years.

Fuel cells may eventually change the way that we think about electric generation. A fuel cell is an electrochemical device that converts a fuel's chemical energy directly to electric energy. Fuel cells have no moving parts and are like a battery except that while batteries only store energy, fuel cells can actually produce electricity continuously given an ongoing supply of fuel. Fuel cells can run on various fuels including natural gas, gasoline, biogas, methanol, ethanol, and hydrogen. The big advantage to fuel cells is that with certain fuels their emissions consist of water and oxygen, and that they are well-suited to CHP applications. Thus they are very environmentally friendly at the point of generation. Some analysts believe that within this decade, stationary fuel cells may become a valid option for widespread use in distributed electric production, thereby fundamentally changing our electric systems over the next fifty years. Lastly, research into development of economic electric storage continues. Technologies showing promise include compressed air storage, new generation batteries and flywheels. If these technologies prove feasible, development of cost-effective storage technologies could greatly reduce peak generation costs.

CARBON CAPTURE AND SEQUESTRATION

Carbon sequestration refers to the capture and storage of carbon as an approach to reducing greenhouse gas emissions during the power generation process. In this process, carbon is removed, usually as carbon dioxide (CO_2), either prior to the combustion of the fossil fuel or after combustion in the exhaust stack. The CO_2 is then transported via pipeline or other means to a location where it can be stored. CO_2 might be stored in deep underground geological formations or in the ocean, or by transforming it into mineral carbonates. Most current research is focusing on geological formations. While carbon sequestration has been proven in concept, it is not yet developed for commercial scale use. And the effects of long-term storage are still unknown. Costs of adding and operating technology for carbon capture and sequestration will certainly increase the cost of power generation. However, this concept may provide a useful tool in the battle to reduce greenhouse emissions from power generation.

What you will learn:

- What transmission is
- Types of transmission
- The physical characteristics of the transmission system
- Operation and planning of the transmission system
- Transmission system costs
- Ownership of transmission systems
- Issues with transmission system construction
- The current status of the transmission grid in the U.S.

5

SECTION FIVE: TRANSMISSION

Electric transmission is the movement of large amounts of electricity over long distances. In this process electricity is moved from a central generating unit to an interconnection with an electrical distribution system, or in some cases, directly to industrial customers. The transmission system is the electrical highway that connects supply to demand across a network called an electric grid. Different entities define the facilities that comprise the transmission system somewhat differently, but transmission generally refers to any electric line with voltage greater than 60 kV (some entities use 40 kV as the break, some 115 kV). Typical transmission voltages include 69, 115, 128, 230, 345, 500, and 765 kV. Transmission lines can be designed to transmit either AC (alternating current) power or DC (direct current) power, but not both. Most lines in the U.S. are AC.

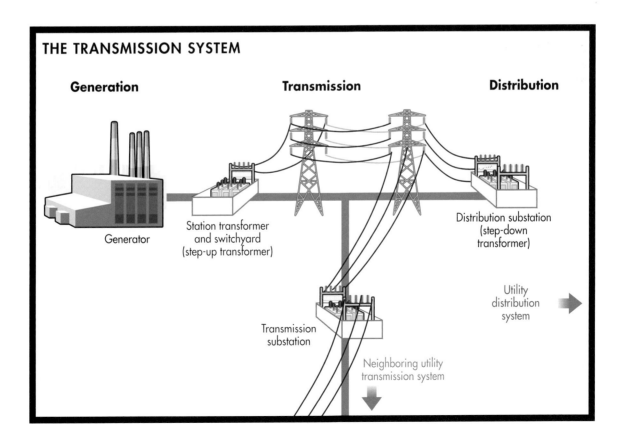

Physical Characteristics of Transmission

In addition to transmission lines, the transmission system includes station transformers, switchyards, and transmission substations. Located outside of the generating unit is a switchyard where the station transformer increases the voltage of the unit output to the voltage of the transmission system. The switchyard also contains switches, breakers, busbars, and other protective equipment that configure the flow of power away from the generating unit and provide protection for both the unit and transmission grid. On the other end of the transmission system is the transmission substation which is the interconnect to another entity's transmission lines. The substation contains switches, breakers and other protective equipment, and, if necessary, transformers to adjust voltage between different incoming power lines. The transmission system connects to a distribution system at the distribution substation, which is generally considered part of the distribution system.

A key consideration in building transmission is to minimize capital and operating costs relative to the transmission capacity. Capacity is largely determined by wire size and voltage. High voltage transmission is the preferable way in which to move bulk amounts of power because higher voltages require lower currents. This reduces not only losses, but also the size of the wire required to provide a specific transmission capacity – thereby reducing necessary capital investment. Transmission lines are generally situated overhead (above ground) except in dense urban areas where underground cables or busbar may be used.

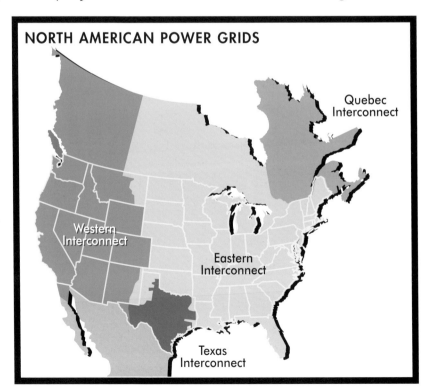

The construction of much of the U.S. transmission network was completed during the 1960s and 1970s. Prior to 1950, maximum transmission voltages were typically 138 kV. From the mid-1950s through the 1960s, 345 and 500 kV systems were built. And by the early

1970s, lines with voltages as high as 745 kV were being added to the transmission system. Construction of transmission lines usually occurs on a large scale since it is more economical to construct transmission lines with larger capacities.

By the late 1970s (after some 30 years of large-scale construction), we had reached a point where significant new construction was no longer required. In fact, there was even excess capacity in some areas. This, coupled with reduced growth in electricity demand and new technical innovations that allowed existing lines to be used more efficiently, allowed the industry to minimize investments in transmission throughout the 1990s – and to maximize the use of those investments already made. In the last ten years, however, demand growth, the growth of wholesale trading and construction of new generation have surpassed transmission capabilities. Recent developments, including the blackout of 2003, a push for renewable energy development and use of locational pricing that makes transmission weaknesses transparent, have led many to call for significant upgrades in our transmission network. And by 2008 we had begun to see what could be a wave of new transmission construction.

Over the years, utilities have interconnected their transmission grids for reliability and supply cost sharing purposes. In the U.S., the transmission system has evolved into three regional grids – the Eastern Interconnect, the Western Interconnect, and the Texas Interconnect. The three regional interconnects, along with the Quebec Province grid in Canada, cover most of the U.S. and Canada as well as border areas of Mexico.

Operation and Planning of the Transmission System

Operation of the transmission system is integrally tied to both customer demand and to the operation of generation resources. Thus the generators and transmission systems must be operated in synch if they are to effectively respond to customer demand. This function is called system operations and is discussed in much greater detail in Section Seven.

An important function of system operations is to operate the transmission system at a capacity that will not harm the system. The optimal capacity of a transmission line is determined by three factors – thermal/current constraints, voltage constraints and system operating constraints. Thermal constraints are limits set to avoid overheating a line. Due to resistance within the line, the flow of electrons causes heat to be produced. The temperature of a given line is determined by variable factors such as current flowing and ambient conditions such as temperature and wind speed (which affect dissipation of heat into the air). These factors can cause the line to overheat, which is dangerous because it can result in sagging (thereby putting the line in contact with trees or other conductors and potentially causing short circuits) and/or permanent damage to lines. To ensure the lines do not overheat, the system operator will set thermal limits on the system. Voltage constraints restrict the voltage that can be carried through a specific line. Excess voltages can result in short circuits, radio interference and/or damage to transformers or customer equipment. And finally, system operating constraints are limitations associated with the need to maintain power flows throughout the transmission and distribution grids. Allowing too much power to flow on a specific line can result in redirection of flows that may interrupt proper operation of the grid.

The combination of transmission system capabilities and generation resources determines the reliability of supply available to a given region. Historically, transmission and generation resources were planned, built and operated on an integrated basis by vertically-integrated utilities. As certain areas of the country have moved into electric restructuring – resulting in the break-up of the vertical utility – operation and planning of the transmission system has fallen to Independent System Operators (ISOs) who operate separate and distinct from generation owners and distribution utilities. ISOs and regulators are working to develop new mechanisms to provide for sufficient transmission planning in a marketplace where generation is developed based on market conditions rather than on integrated plans associated with load growth.

Transmission System Costs

The bulk of costs associated with transmission service are the initial capital costs to build the lines. A lesser cost is the annual maintenance associated with keeping the lines operating reliably. Costs to build transmission lines vary from about $130,000 per mile for lower voltages (115 kV) to $500,000 per mile for higher voltages (230 kV) to as much as $1,000,000 per mile for ultra high voltages (345 kV and above). The capacity of each line is also an important factor when considering cost. Underground transmission is significantly more expensive, often costing as much as ten times more per mile than overhead.

Ownership of Transmission

Until the advent of electric restructuring in the U.S., transmission lines were owned by vertically-integrated utilities or by federal generation agencies. As restructured electric markets have evolved, some utilities have concluded that it no longer makes sense to own transmission lines, and have sold theirs to transmission companies (also referred to as transcos). A transco is a stand-alone owner and operator of transmission facilities. Many industry observers believe that over time we will evolve to a market structure where transmission ownership and operation is dominated by transcos. This is similar in structure to the current U.S. interstate natural gas transmission system, where investor-owned companies own and operate interstate pipelines as a stand-alone business.

Issues with Transmission Construction

Most experts agree that the U.S. electricity infrastructure needs considerable attention. Yet building new transmission lines has proven to be difficult in recent years. This is due to a number of factors that include public opposition, regulatory issues and financial uncertainties. Public concerns about new transmission lines include land use issues, impacts on property values, environmental issues, and perceived electromagnetic field (EMF) threats. Local opposition is difficult to overcome when impacts may fall locally, but benefits apply regionally.

Regulatory issues have also impeded new transmission construction. Key issues include mechanisms for recovery of costs and profits in rates, how costs are assigned to market participants, and procedures for obtaining construction permits. Rates for transmission lines used for wholesale commerce are set by FERC. As deregulation of wholesale markets moved forward in much of the United States in the late 1990s and early 2000s, uncertainty over exactly how costs would be recovered slowed down investment in transmission. FERC responded by attempting to set specific principles for transmission rates and offering higher rates of return for new transmission construction. The issue of how transmission costs should be allocated persists, especially in ISO markets where they are potentially spread across a broad geographical area. It is not uncommon for participants in one region to oppose a project that they believe only benefits another region. Lastly, permits for construction are issued at the state level. This means that projects crossing multiple states must obtain permits from each state in a separate proceeding. Again, local interests may trump wider regional interests making it difficult to get all necessary permits. The Energy Policy Act of 2005 enacted a mechanism that was intended to allow project proponents to appeal to the federal government if states blocked permits, but as of late 2009 this provision has yet to be tested.

SECTION FIVE: TRANSMISSION

The Current Status of the U.S. Transmission System

Historically, expenditures for new transmission construction in the U.S. have tracked growth in both customer demand and generation capacity additions. In recent years, however, investment in transmission has not kept up with either demand growth or growth in generation construction. In areas where market structure has moved to one of non-utility generation construction, the traditional integrated generation/transmission planning has been eliminated. Restructuring has also led to increasing use of the transmission grid for something it was not designed for – to move electricity within a large region based on a competitive wholesale trading marketplace. The inevitable result has been increasing issues with transmission adequacy and greater need for transmission operators to manage the use of the transmission grid through transmission congestion procedures. This is clearly demonstrated by the increasing number of incidents that have caused system operators to implement transmission loading relief (TLR) procedures (A TLR incident reflects a situation when Reliability Coordinators are required to implement procedures to mitigate potential or actual operating limit violations). The graphic below shows the increasing prevalence of these incidents.

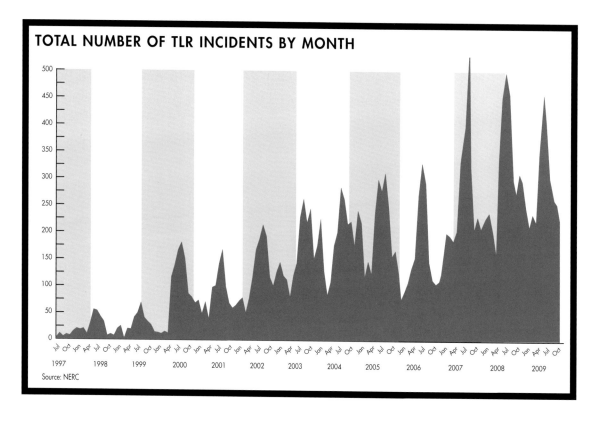

So we return to the inevitable conclusion that the current transmission grid is inadequate to serve our ever-growing needs. Given that regulators and consumers see the same issues, utilities and merchant transmission companies have begun the process of developing new projects. Their efforts are spurred by support from the federal government, including authorization for higher rates of return for certain new transmission projects. NERC estimates that by 2013 U.S. transmission will have been expanded by more than 5% over 2008 levels. But the barriers to siting and building new projects are still strong, and it remains to be seen how many projects will actually get built. Thus transmission is one area of the electricity industry where uncertainty abounds. We may, in fact, see a new wave of construction. We may also be far from seeing what so many have called for. Or, significant enhancement of our transmission system may await commercialization of new technologies that will allow existing lines to be upgraded to carry more power.

What you will learn:

- What distribution is
- Types of distribution
- The physical characteristics of the distribution system
- Operation of the distribution system
- Distribution system costs
- Ownership of distribution systems
- The current status of distribution systems in the U.S.

SECTION SIX: DISTRIBUTION

Electric distribution is the movement of electricity from the interconnection with the transmission system through the end-use consumer's meter. If transmission is considered the highway on which electricity travels long distances, the distribution system can be considered the streets and avenues that connect end-use customers to it. Generally, distribution refers to electric systems with voltages lower than 60 kV (although some utilities define distribution as lower than 40 kV). Distribution systems are often divided into primary systems (higher voltages) and secondary systems (lower voltages). Ultimately, the voltage at which electricity is delivered to an end-use consumer must be transformed to the voltage used by the consumer's electrical devices, which for smaller customers is the common 120 V we are used to seeing in our homes and offices. All distribution lines in the United States distribute AC power.

Physical Characteristics of Distribution

In addition to distribution lines, distribution systems consist of transformers, voltage regulators, switches, circuit breakers, automatic reclosers, power capacitors, monitoring systems, service drops, and customer meters. As mentioned above, the distribution system is generally divided into two categories – primary distribution (which is most commonly at voltages of 12.4 or 13.8 kV), and secondary distribution (which is most commonly at 120, 240, or 480 V). Primary voltages are used to move electricity throughout the utility distribution area since distributing power at higher voltages results in less significant line losses. And secondary voltages are used as the lines approach a group of customers. Some industrial and large commercial customers take service at primary voltages (and in the case of very large industrial customers, even at transmission voltages). However, most customers – because they do not have the equipment necessary to transform it themselves – take voltage at the level

COMMON VOLTAGES

Primary Distribution
- 4,160
- 6,900
- 12,470
- 13,200
- 13,800
- 34,500

Secondary Distribution
- 120
- 240
- 480

SECTION SIX: DISTRIBUTION

THE DISTRIBUTION SYSTEM

required by their appliances. In residential and rural areas, the most common supply voltages are 120/240, while in commercial or industrial areas service voltages are 208/120 and 480/277 (dual voltages refer to the different voltages that are available at one service drop).

Power from the transmission grid enters the distribution system at the distribution substation. The distribution substation consists of transformers that step-down voltages from transmission levels to primary distribution levels. After it is transformed, electricity leaves the distribution substation at the bus bar, which is a large piece of metal conductor that allows multiple circuit connections. From the distribution substation, electricity is then distributed through the utility's service area in primary feeders. These carry the electricity to clusters of end-use customers and are also used to directly serve large customers who have their own transformers within their facilities. In areas with numerous smaller customers, step-down transformers reduce the voltage to secondary levels where it can be used by consumers. Depending on the voltage of the service line, a final transformer is often required outside the customer premises to provide the necessary service voltage. These are the transformers seen on poles in overhead areas, the pad-mount transformers seen as green cabinets (often outside commercial businesses), and the underground vault transformers located next to residential subdivision sidewalks. Distribution lines may be either overhead or underground. Overhead lines are cheaper to build and easier to fix, but are more exposed to

UNDERSTANDING TODAY'S ELECTRICITY BUSINESS

> ## VARs AND POWER FACTOR
>
> A primary use of electricity is to drive electric motors. Electric motors create the mechanical energy of a spinning shaft by using electrons to create a magnetic field which causes the shaft to spin. The electricity that magnetizes the coils does no work, and due to the physical nature of electricity, does not make the electric meter turn. Thus a portion of the electric current that a generator and electrical system must provide does no work and is not recorded by a meter. This is called reactive power, which is measured in units of volt-ampere reactive, or VAR. The portion of electric current that does work and turns the meter is called real power, which is measured in units of watts, kilowatts, etc.
>
> The overall power which the system must be designed to deliver is called apparent power and is the sum of reactive power plus real power
>
> Apparent power = reactive power (VAR) + real power (Watts)
>
> Power factor measures the relationship of real power to apparent power:
>
> $$\text{Power Factor} = \frac{\text{Real Power}}{\text{Apparent Power}} = \frac{\text{Real Power}}{\text{Reactive Power} + \text{Real Power}}$$
>
> VARs are produced by certain types of generators and can also be produced by other equipment such as capacitor banks placed on the distribution system. Since the utility needs to be paid for all the services it delivers, it prefers that customers have power factors as close to 1.0 as possible (since any reactive power is not metered). For industrial customers, utilities will often measure the power factor at the meter, and will bill industrial customers for power factor deviations outside of acceptable ranges. If these charges get too high, customers can install equipment on their side of the meter to better manage their power factor.

potential hazards and in some areas are considered unsightly (i.e, unsightly enough to warrant the extra expense to bury them). Underground lines are more expensive to build and fix, but require less maintenance and do not cause visual issues.

A key consideration for distribution lines is that they carry electricity safely and in a way that will not damage customer equipment. Thus the system is designed to quickly isolate short-circuits and to maintain proper power quality (frequency and voltage). Voltage regulators ensure that system voltages remain within acceptable limits. Switches allow various circuits to be switched or connected to other circuits in different configurations. Circuit breakers are mechanical devices allowing circuits to be isolated (taken off the grid) so that maintenance can be performed and/or to protect the system in the event of power quality problems or circuit overloads. Automatic reclosers are circuit breakers that automatically interrupt circuits when a fault is sensed. Automatic reclosers give temporary faults repeated chances to clear themselves by automatically closing a few times (often three times) before finally locking open and isolating the circuit if the fault fails to clear. These are commonly used throughout distribution lines as the first line of defense in minimizing power problems. Power capacitors are used in distribution systems to supply reactive volt-amperes (VARs) to the system (see box for explanation of VARs). At

SECTION SIX: DISTRIBUTION

each customer location, a service run or drop consists of the cabling and protective equipment that connects the distribution system to the customer's internal wiring. And, of course, each utility must maintain a meter to measure the amount of energy used.

To ensure that it runs smoothly, the distribution facilities are monitored remotely using a SCADA (Supervisory Control and Data Acquisition) system. Typically SCADA systems are installed at concentrated points such as major substations so that the distribution operations center can monitor loads, status of circuit breakers, voltage, and VARs. Historically it has not been considered cost-effective to install SCADA systems throughout much of the distribution system, but as costs decline, more and more of the system is now remotely monitored and even controlled.

Distribution circuits are designed with specific customer loads in mind. Thus the size and type of circuit will be engineered based on the size and types of loads expected to be served. The addition of a major new customer, or change in existing customer loads can often require reconfiguration or upgrading of a distribution circuit. Different types of distribution systems include radial feed, loop feed and network system and are discussed in greater detail below.

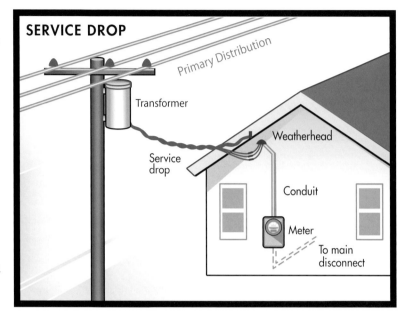

SERVICE CONFIGURATIONS

Electrical services are provided in different configurations depending on the needs of the customers. Common variables for service are whether the service is single phase or three-phase and whether the service is 2-wire, 3-wire, or 4-wire. Single-phase service is sufficient to operate typical lighting and small appliances and is generally what is provided to residential and many rural customers. Three-phase service is preferable where large motor loads exist and is commonly provided in commercial and industrial areas.

The number of wires refers to how many conductors are connected to the service. Electricians who wire buildings provide voltages to different types of electrical equipment by wiring the equipment to the conductors in different configurations. For instance, a 208/120 service would provide 120 V single-phase power for use by appliances and light bulbs, while also providing for 208 three-phase service for motors.

UNDERSTANDING TODAY'S ELECTRICITY BUSINESS

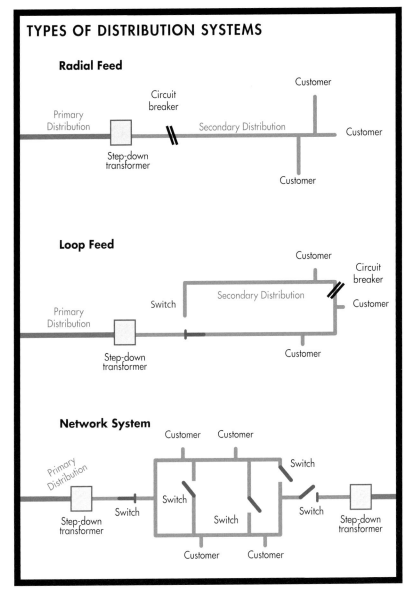

Radial Feed

A radial feed is simply a single line from a transformer out to a number of customers. While the lowest cost of the options, radial feeds do suffer from the fact that loss of cable, primary supply or the transformer will result in loss of service to all customers on that feed. Also, radial circuits must be de-energized to perform routine maintenance and services.

Loop Feed

A loop feed serves customers off a loop that is connected to the primary feed at two ends. This costs more than a radial system since duplicative equipment is required, but does provide the capability of isolating faults within the loop and continuing to feed all customers except those on the section with the fault. The reliability of both radial feeds and loop feeds can be enhanced by adding additional primary distribution feeds to the circuit.

Network System

A network system connects multiple primary feeds and interconnects multiple distribution circuits in the form of a grid. While network systems provide the highest form of reliability since customers can be served in multiple ways, they are also expensive because of the costs of duplicative equipment, transformers and specialized network protective equipment. Networks are generally used in downtown urban areas with highly dense critical loads.

SECTION SIX: DISTRIBUTION

Operating and Planning of the Distribution System

Day-to-day operation of distribution systems is the responsibility of the distribution dispatch or distribution operations center. The dispatch center monitors flows and other system status values at points where SCADA systems have been installed. The dispatch center also directs system maintenance activities and oversees responses to any system outages or disturbances. During outages or disturbances, the dispatch center will direct crews on switching of circuits to restore power where possible and to ensure that lines are de-energized where crews are working. The dispatch center then restores the system to its normal operating configuration once appropriate repairs have been performed.

Maintenance management is a longer-term, but similarly important function of operating a distribution system. Periodic routine maintenance as well as unplanned maintenance in response to system conditions must be performed to keep systems operating properly. A specific planned maintenance program is critical to prevent both deterioration of equipment as well as vegetation from growing into power lines. Identification of the equipment that is critical to safe operation of the system and developing contingency plans for its potential failure is also important.

Lastly, system planning studies are conducted periodically to determine the need for enhanced maintenance or system expansion. System models can be used to determine optimized operations and configurations, and overall system reliability and efficiency can be enhanced by frequently re-evaluating circuit design. And, of course, customer loads continually come and go, so system planners are always engaged in evaluating the need for system modifications to meet changing customer needs.

> **THE METER**
>
> The last key component of the distribution system is the meter located at each customer location. Without the meter, customers cannot be billed and energy companies cannot be paid. Metering is currently undergoing significant transition. Until the last 15 years, almost all meters were read once a month by a meter reader who recorded usage at each customer location. Meter data was generally limited to kWh used and for larger customers maximum kW. Since then we have become accustomed to increasingly sophisticated meters at reduced costs. Now many utilities depend on meters that can be read remotely and make meter data available on a real-time basis. Higher-end meters can be used to record large amounts of useful data including energy usage by time period, demand by time period and various measures of power quality. In the next few years, the electric meter may evolve into a services gateway that will allow two-way communication between energy providers and consumers, opening up new ways for energy companies to maximize the efficiency of supply and customers to participate in energy markets.

Distribution System Costs

Because distribution systems are designed specifically to meet customer needs and no two distribution systems are the same, there is no rule of thumb concerning their cost. However, we can say that the portion of a customer's rate that reflects distribution costs is higher than the transmission component. It is also more costly to serve smaller customers than to serve larger customers since smaller customers are served at lower voltages and thus require more equipment. And it is significantly more costly on a per-customer basis to serve customers in either dense urban areas (need for underground service) or remote rural areas (need for a long distribution line to serve a small number of customers).

Ownership and the Current Status of Distribution Systems

Most distribution systems are owned and operated by a distribution utility. In some limited cases distribution systems are owned and operated by private entities such as military bases, large industrial complexes or private developments. Distribution utilities can be divided into three categories – investor-owned utilities, municipal utilities (and their close cousins public utility districts) and rural electric co-ops. Investor-owned utilities, or IOUs, are for-profit companies owned by shareholders whose distribution functions are highly regulated by the state utilities commission. Municipal utilities and public utility districts are owned by local government entities, most commonly a city, and are usually not regulated by the state. Rural electric co-ops, also known as rural electric agencies (or REAs) are common in rural areas and are owned by their ratepayers and run by an elected board.

Distribution systems have been less affected by electric restructuring in the United States than have the generation and transmission sectors. An exception, however, is that many IOUs have been affected in regions where the vertical integration of generation, transmission and distribution functions has been broken apart. This has forced the IOUs to restructure their business operations to create stand-alone distribution organizations. And in areas where customer access to alternative power suppliers has been allowed, the distribution function has been further restructured by elimination of the distribution utility responsibility for at least a portion of the supply function. Restructuring has generally not been applied to either municipal utilities or REAs, and most still operate as vertical utilities.

SECTION SIX: DISTRIBUTION

The Smart Grid

As of the late 2000s, many electric utilities had begun a transformation to what is known as the smart grid. While this transformation affects both transmission and distribution sectors, the effect will likely be greater on distribution since some of the components of the smart grid have been in use in the transmission sector for many years. In fact, since many of the smart grid technologies have already been in use in portions of the electric system, the transformation might be better described as moving from a somewhat smart grid to a smarter grid.

Various parties use the term smart grid to define different related technologies and operating practices. But generally, the term applies to deployment of various digital technologies at points throughout the transmission and distribution grid, and then implementing internal procedures and external services that make use of the capabilities of the new technology. The technologies include integrated two-way communication technologies, advanced control methods, interconnected monitoring and metering equipment, advanced electronic grid components, decision support systems, and human interfaces. Key activities the technologies may enable include two-way communication with system components and with customers, real-time collection and presentation of system and customer data, improved grid modeling and real-time diagnostics, remote control of system and customer devices, automatic response to abnormal system conditions, information-based scheduling of maintenance, automatic outage detection and response, remote turn-on/turn-off of customers, remote meter reading, enhanced energy theft detection, integration of distributed storage and possibly electric vehicles, and implementation of advanced demand response and distributed generation services.

The extent and pace of the roll-out of smart grid technologies is uncertain as of late 2009. Federal subsidies have encouraged many utilities to move forward with implementation plans. But concerns that may slow down development do exist. Perhaps the biggest question is will the benefits offset the costs or will the smart grid result in overall net rate increases thus causing a backlash from consumers. Other issues include whether smart grid technologies cause cyber security risks by allowing pathways to access grid systems and whether allowing the smart grid to collect and disseminate consumer data creates privacy concerns.

What you will learn:

- Operational characteristics of power systems
- The role of system operations
- Who is responsible for systems operations
- How system operations schedules generation, reserves and transmission
- How supply and demand are matched in real time
- How the role of system operations is changing

SECTION SEVEN: ELECTRIC SYSTEM OPERATIONS

The physical electric system is comprised of a highly complex and interdependent network of generators, transmission and distribution systems, and customer loads spanning thousands of miles. Despite the diversity of the system, it must always be tightly controlled. Supply and demand must be kept in balance continuously and voltages and frequencies must be kept within tight bounds or serious consequences may ensue – the customers' equipment may fail to run properly, grid equipment may be damaged, and, in the worst of scenarios, the system could crash and customers would experience outages. Managing this complex system in the short term (next day) and real time (current hour) is the responsibility of the electric system operator. System operators must schedule resources to ensure supply is available to match demand and, in real time, must continuously adjust generation levels to match demand fluctuations.

Operational Characteristics of Power Systems

Operating power systems is a complex task largely due to a few key physical characteristics. First, electricity cannot effectively be stored. This means that the supply provided by electric generation must be continuously in balance with the demand of customers' usage. The operation of the system is further complicated by the fact that the path of electric flow is very difficult to control. Electrons simply flow on the path of least resistance, whether or not that path matches contractual agreements or the desires of the system operator. Thus interconnected utilities are inextricably entwined with the actions of their neighbors. The only way to avoid this interdependence is for utilities to isolate their systems. But because utilities depend on connections with each other for reliability and access to economic generation sources, this is not a viable solution.

Further complicating system operations is the speed at which system disturbances travel. Changes in voltage or frequency on electrical lines travel at the speed of light, and any major system disturbances can be propagated across an interconnected grid in a matter of seconds. Disturbances can be dangerous because uniform voltages and frequencies must be maintained within strict limits to avoid degrading service to customers (e.g., voltage spikes knock off computers, low voltages dim lights, high frequency speeds the operation of electrical machinery and can damage generating equipment, etc.). And in an information society dependent upon computers and microchips, even momentary outages of computer-controlled equipment can cost customers millions of dollars and are not considered acceptable.

> **KEY CHARACTERISTICS OF POWER SYSTEMS**
>
> - Electricity cannot be stored economically.
> - Supply and demand must always be in balance.
> - The path of electric flow cannot be controlled.
> - Disturbances travel very quickly.
> - Voltages or frequencies outside of limits damage equipment.
> - Even momentary outages are not acceptable.

What System Operations Does

System operators, also known as control area operators or balancing authorities, manage the actions of generators, transmission owners and load serving entities within their designated control area, and coordinate with neighboring control areas and regional system operators to maintain acceptable levels of service. To do this, system operations must:

- Forecast demand in the day-ahead.
- Schedule generation to match forecasted demand.
- Schedule reserves and other ancillary services.
- Schedule use of the transmission system among various market participants.
- Communicate schedules to neighboring operators so flows across interconnections can be anticipated.
- Manage the system in real time by correcting imbalances minute-by-minute.
- Correct any system disturbances that may occur.
- Restore power should an outage occur.

Who Handles System Operations

The U.S. and Canada have four regional grids made up of numerous interconnected transmission systems. These are the Eastern, Western, Texas, and Quebec

Interconnects (see map on page 56), and they generally operate independently of each other. Within each interconnect, however, utilities, transmission owners and generators are tied together in a linked network so that the actions of any one system operator can have strong impacts on the others within their grid. Each interconnect is divided into various control areas. And for each control area there is a system operator that is responsible for system operations. System operators may be vertically-integrated utilities, municipal utilities, federal power agencies, groups of utilities called power pools, or Independent System Operators (ISOs). ISOs are also known as Regional Transmission Operators (RTOs). Smaller utilities without a big enough system to warrant the cost of their own system operations often contract for the service with a neighboring larger utility, so control areas are often made up of multiple utilities. Some utilities have also banded together in groups called power pools that allow a pool of generators to be shared among utilities in order to optimize economic dispatch. In areas where generation markets have been restructured, the system operator function is taken over by the ISO (this has occurred to date in California, Texas, New York, New England, and much of the mid-Atlantic and Midwestern states) and the control areas become regional. Today we have over 100 control areas that vary significantly in size. The smaller control areas manage less than 100 MW of generation while the PJM ISO manages over 165,000 MW of generation.

System operators coordinate their actions through Regional Reliability Councils of the North American Electric Reliability Corporation (NERC). NERC's mission is to ensure that the electrical transmission grid in North America is reliable, adequate and secure. NERC is an umbrella organization made up of eight Regional Reliability Councils. The entities that belong to the reliability councils account for virtually all of the electricity supplied in the U.S., Canada and a portion of Baja California Norte in Mexico. NERC sets forth general standards for operations to ensure reliability, and each reliability council then adapts these standards for its own system operators to follow. Historically dominated by utilities, the regional councils opened their memberships to independent power producers and power marketers in 1994. Members now include transmission owners, ISOs, load serving entities, utilities, merchant generators, marketers, end-users, and government agencies.

The Energy Policy Act of 2005 gave the Federal Energy Regulatory Commission (FERC) jurisdiction over reliability standards with the intent of changing them from voluntary to mandatory. In 2006, FERC certified NERC as the Electric Reliability Organization (ERO) responsible for developing and enforcing mandatory electric reliability standards under the Commission's oversight. In 2007, NERC began the process of filing reliability standards at FERC and many of these were approved late in the year.

SECTION SEVEN: ELECTRIC SYSTEM OPERATIONS

SYSTEM OPERATORS IN THE U.S. AND CANADA

Reliability standards set forth by the regional councils require control areas to maintain a balance between generation and load under normal conditions, re-establish the generation-load balance within 15 minutes of unexpected failure of a generator or transmission line, maintain frequencies and voltages within specific bounds, maintain generation reserves to respond to any disturbances, and avoid overloading transmission lines.

These reliability standards require system operators to plan not only for forecast demand needs plus a reserve factor to cover mis-forecasting, but also for contingencies for outages of system components (power plants or transmission lines). Thus the standards require control areas to maintain a certain amount of available reserves equal to the greater of:

- A percentage of forecasted loads, or
- The most severe single contingency of generation or transmission loss.

UNDERSTANDING TODAY'S ELECTRICITY BUSINESS

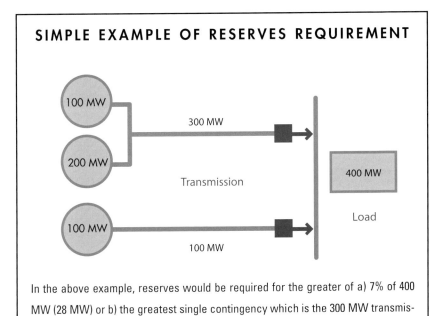

SIMPLE EXAMPLE OF RESERVES REQUIREMENT

In the above example, reserves would be required for the greater of a) 7% of 400 MW (28 MW) or b) the greatest single contingency which is the 300 MW transmission line. So in this case, the system would need 300 MW of reserves.

These reserves must be available to quickly bring supply and demand back into balance should there be an unexpected outage of either generation or transmission.

Forecasting and Scheduling

Scheduling generation and the transmission system is a complex task. To begin this process, system demand must first be forecast. Then the system operator matches available supply to meet the forecasted demand, taking into consideration system constraints as well as the need for reserves should anything unplanned occur. On the following page are brief descriptions of this two-part process.

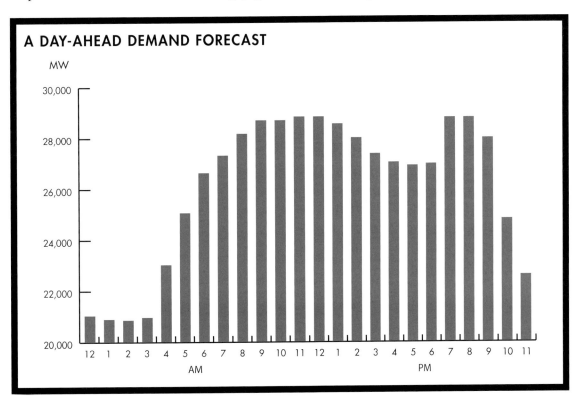

A DAY-AHEAD DEMAND FORECAST

77

SECTION SEVEN: ELECTRIC SYSTEM OPERATIONS

Demand Forecasting

Demand is forecast using models developed from historical demands given forecasted weather patterns and business activity. Using these models, system operators develop hour-by-hour day-ahead demand forecasts. Demand models are rerun during the day of delivery so that forecasts can be continually adjusted based on changes in weather or other factors.

Scheduling Generation, Transmission and Reserves

Once the demand forecast is set, system operators then look at available generation – both from within the system as well as any generation that might be available from neighboring systems. System operators schedule available generation on an hour-by-hour basis to match load and reserve requirements. Their goal is to optimize the schedule to a) ensure reliability standards are met and b) minimize overall supply costs (this is called least-cost dispatch subject to constraints or scheduling optimization). Variables that must be considered in scheduling optimization include the ability of the transmission grid to move power to various parts of the system, generation that must be run to keep the system operating safely, and environmental and regulatory requirements. The physical ability of the system to serve all customers given a proposed schedule is determined by a power flow model. If the power flow model results in a non-feasible schedule, then schedules must be adjusted until feasible solutions are found. Generation that must be running to support the grid is called must-run generation, while generation that must be run for environmental and regulatory reasons (for instance, the requirement to take power from QFs) is called must-take generation.

Generation and transmission schedules are set forth on an hour-by-hour basis in the day-ahead, and generating units are notified of their schedules. Units may be scheduled to provide energy (meaning they will be running) or to provide reserves (meaning they are ready to run as needed). In some cases, units may be scheduled so that a portion of the capacity is used for energy and the rest is used for reserves – for example a 400 MW unit might be scheduled to provide 300 MW of energy and 100 MW of reserves. On the day of delivery, unit schedules may be adjusted in the hour-ahead based on changes in the forecast.

Ancillary Services

Ancillary services refers to the services (other than energy) required by system operators to ensure safe and secure operation of the electric grid. Ancillary services include:

SIMPLE EXAMPLE OF SCHEDULING

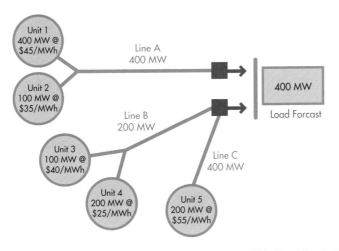

In this simple model, system operations has forecast 400 MW of load for the hour. In attempting to optimize scheduling (lowest to highest cost generation), the system operator would prefer to schedule:

Unit 4	$25	200 MW
Unit 2	$35	100 MW
Unit 3	$40	100 MW

Unfortunately, this is clearly not feasible given the limited capacity of transmission line B (only 200 MW). So the optimized dispatch, subject to constraints will be:

Generation Schedule

Unit 4	$25	200 MW
Unit 2	$35	100 MW
Unit 1	$45	100 MW

Transmission Schedule

Line A	200 MW
Line B	200 MW

In addition to the units scheduled for energy, the operator will also need reserves. Since the greatest single loss contingency is 200 MW on Line B, the operator will need to schedule 200 MW of reserves. The safest place to obtain the reserves is from Unit 5, since the loss of Line A would create a situation where reserves from Unit 1 would not be available to the system. If Unit 5 is scheduled for reserves, this would also necessitate scheduling Line C for 200 MW to ensure the transmission capacity is available if supply from Unit 5 is needed.

Automatic Generation Control (AGC)

Also called regulation, AGC is used to manage the minute-by-minute fluctuations in system loads. AGC units can be ramped up and down remotely from the system operations floor and are used throughout the hour to keep supply and demand in balance. Typical sources of AGC include hydro power, gas combustion turbines, and natural gas steam-turbine units that are providing energy from a portion of their capacity, but that have additional unused capacity.

Spinning Reserves

Spinning reserves refer to units that are already synchronized to the frequency of the system, and thus can begin providing energy to the grid upon receiving a dispatch call. Capacity included in spinning reserves must be fully available to the system operator within ten minutes of notification.

SECTION SEVEN: ELECTRIC SYSTEM OPERATIONS

Typical sources of spinning reserves include hydro power, gas combustion turbines, and natural gas or coal steam-turbine units that are providing energy from a portion of their capacity, but that have additional unused capacity.

Non-spinning Reserves

Non-spinning reserves are units that are not synchronized to the frequency of the system, but can be available within ten minutes of notification. Non-spinning reserves can also include demand response that is available within the ten minute window. While non-spinning reserves have the same ten minute requirement as spinning reserves, these units take longer to begin contributing partial generating capacity since they must first be synchronized to the system. Typical sources of non-spinning reserves are similar to those used for spinning reserves.

Supplemental Reserves

Supplemental reserves are units that are available with a longer lead time, often thirty minutes from notification. Typical sources for supplemental reserves are coal and gas steam-turbine units that already have warm boilers.

Voltage Support

Voltage support is provided by specially equipped units that have the capability to provide VARs to the system (for an explanation of VARs, see page 65).

Black Start

Most units cannot start up without electricity from the grid. This causes a problem for restoring the grid if an outage has occurred. Black start units can start independently without electricity from the grid. System operators need to maintain a certain amount of black start to ensure their ability to restore the grid should there be an outage.

Once the units have been scheduled to provide ancillary services in the day-ahead, the plant operators are told how they are expected to operate each hour of the day for the following day. However, these schedules may be adjusted in the hour-ahead (actually two hours prior to operation) if forecasts or unit or transmission line availability change.

How Supply and Demand are Kept in Balance in Real Time

After the system operator has forecast demand, scheduled the system in the day-ahead to meet the forecasted demand, and then adjusted that schedule in the hour-ahead, he

CASCADING OUTAGES

Blackouts generally begin when a large transmission line or generating unit suddenly drops off-line due to an unexpected event. Examples include power lines coming in contact with trees, lightning strikes or equipment failures. Normally, power system operators are able to compensate for the loss of one source of supply by rapidly bringing on reserve sources so that supply and demand stay in balance. But if the system is already running tight due to a prior incident, or if system response is inadequate due to physical circumstances, equipment failure and/or operator error, additional system disturbances may occur. As supply and demand get out of balance in a certain region, voltage and frequency spikes begin to propagate through transmission lines. Because these spikes can be damaging to equipment, transmission lines and generating units have automatic or manual relays that are designed to island troubled areas from the rest of the interconnected grid. If islanding is done quickly, and surrounding areas have adequate generation to support their demand, blackouts can be limited to a small area. However, if efforts to island the disturbance fail, the problems will continue to cascade throughout the interconnected grid.

Although rare, cascading blackouts do occur. The most recent large scale event was the blackout on August 14, 2003, which affected over 50 million people in the Midwest, Northeast and Ontario. Following are details of some of the major outages over the last 40 years:

Date	Locations Affected	Customers Affected	Duration
November 9, 1965	Virtually all of NY, Connecticut, Massachusetts, Rhode Island, and much of Ontario	30,000,000 customers, 20,000 MW of demand	Up to 13 hours
July 13, 1977	New York City	9,000,000 customers, 6,000 MW of demand	Up to 26 hours
July 2, 1996	Arizona, California, Colorado, Idaho, Montana, Nebraska, Nevada, New Mexico, Oregon, South Dakota, parts of Texas, Utah, Washington, Wyoming, Alberta, British Columbia, and Baja Norte	2,000,000 customers, 11,850 MW of demand	From a few minutes to several hours
August 10, 1996	Arizona, California, Colorado, Idaho, Montana, Nebraska, Nevada, New Mexico, Oregon, South Dakota, parts of Texas, Utah, Washington, Wyoming, Alberta, British Columbia, and Baja Norte	7,500,000 customers, 28,000 MW of demand	Up to 9 hours
June 25, 1998	Minnesota, Montana, North Dakota, Wisconsin, Ontario, Manitoba, and Saskatchewan	152,000 customers, 950 MW of demand	19 hours
August 14, 2003	Parts of Ohio, New York, New Jersey, Michigan, Pennsylvania, Connecticut, Massachusetts, Vermont, and Ontario	50 million customers, 61,800 MW of demand	Up to 2 days, rolling blackouts in Ontario for up to one week

You may remember that California instituted rolling blackouts during the California energy crisis of 2000-2001. These blackouts were not the result of a cascading outage, but were planned load reductions to keep demand in balance with limited electric supply.

must then monitor and manage the system in real time. During the hour, the system operator will normally keep the system in balance by ramping AGC units up and down in response to load fluctuations. If loads are higher than forecast, or sources of supply are lost unexpectedly and AGC is not sufficient, then the operator will call on spinning reserves to keep the system in balance. This, however, leaves the operator short on spinning reserves should another unexpected event occur. It is likely that he will then move non-spinning reserves to spinning status, or move the non-spinning reserves directly on-line as energy, and then ramp the spinning units back down once the additional generation from non-spinning resources becomes available. In this way the operator can maintain the system within the NERC criteria even during contingency events.

Occasionally, despite the best efforts of system operators, the system will get out of balance. The result is voltages and/or frequencies outside the accepted limits. High voltages or frequencies are managed by reducing supply, while low voltages or frequencies require increasing supply. The last line of defense against low voltages or frequencies is to take loads off-line to get supply and demand back in balance. Many utilities have interruptible customers whose supply can be interrupted for reliability reasons. As a last resort, the system operator must involuntarily interrupt customers by isolating their circuit from the grid. This is called a rolling blackout.

The Changing Role of System Operations

In this section, we have described the physical actions that a system operator must take to ensure reliable operations of the grid. We have not discussed the market structures that are necessary to make resources available to the system operator, nor the commercial mechanisms for determining how those resources should be scheduled and paid for providing services. Historically, system operations has been handled by a vertically-integrated utility or a power pool that has direct control over all the units on its system. In this simple structure, units are scheduled based on marginal costs of operation and any system constraints. As parts of the U.S. have moved into competitive electric markets, the necessary commercial arrangements have become much more complex. Now system operators are asked to balance supply and demand when demand is served by numerous competing entities and generating units are owned by a multitude of profit-driven merchants. Keeping the system working requires a carefully crafted set of market arrangements. Different market structures have been tried – some have worked well and others have failed notably. In Section Nine we will address the various market structures and the related trading arrangements and rules for scheduling systems in competitive environments. But first a look at the various market participants in the vertically-integrated and restructured models.

What you will learn:

- Who market participants in the traditional vertically-integrated marketplace are

- Who market participants in the competitive market models are

- The roles of the various market participants

8

SECTION EIGHT: MARKET PARTICIPANTS IN THE DELIVERY CHAIN

Because the different market structures found in the U.S. alter both the makeup and roles of market participants, we will need to discuss these entities twice. We'll first discuss participants in the traditional vertically-integrated utility market model, then we'll look at the various market participants in the more complex competitive models. We will discuss the market models themselves in the following section.

Participants in the Vertically-Integrated Market Model

The traditional U.S. market model is simply a vertically-integrated monopoly that handles all electric service functions as an integrated entity. This means the utility owns the generation and transmission necessary to serve its end-use customers, manages system operations to serve them, and is the only entity providing electric distribution and supply. Different types of integrated monopoly utilities include investor-owned utilities, munis and co-ops.

Investor-Owned Utilities

Investor-owned utilities, or IOUs, are for-profit corporations owned by either public or private shareholders. Most are publicly owned and their stocks trade on Wall Street. Because these entities are for-profit, they must be regulated to ensure that the interests of consumers are being preserved. Each IOU is assigned a specific franchise service territory and is responsible for serving all electric consumers within that area. No other entity is allowed to provide electricity services in the IOU's service territory. Traditionally, IOUs owned all their own generation, transmission, and distribution and managed their own systems operations. Due to a federal regulation called PUHCA that was only recently repealed, IOUs were mostly prohibited from crossing state lines and thus a fragmented market of numerous IOUs evolved. IOUs typically interconnect, however, and trade with each other to take advantage of cost and reliability benefits. The U.S. has 212 IOUs that serve about 70% of the customers in the U.S.[1]

---

[1] Data from the American Public Power Association 2009-10 Annual Directory and Statistical Report.

SECTION EIGHT: MARKET PARTICIPANTS IN THE DELIVERY CHAIN

LARGEST IOUs IN THE U.S.	
	1,000 MWh
1. Florida Power and Light	105,275
2. Georgia Power	86,084
3. Southern California Edison	79,505
4. Pacific Gas and Electric	79,451
5. Virginia Electric and Power	75,631
6. Duke Energy Corp.	57,009
7. Alabama Power	56,642
8. Detroit Edison	48,816
9. Commonwealth Edison	48,557
10. Florida Power Corp.	39,282
2007 data from EIA	

Municipal Utilities and Public Utility Districts

In some areas, local governments are responsible for providing electric services rather than allowing for-profit IOUs to handle them. In many cases, these utilities are run by the city government and are called municipal utilities, or munis. In certain states, these utilities are run by a group of cities or a county, and are called public utility districts, or PUDs. Munis and PUDs are non-profit organizations that are run by a local government agency. Municipal utilities generally operate as a division of the local city government and provide electricity in the same way that many cities provide water, sewer, garbage, and other utility services. In most states, they are not regulated by the state. Larger municipal utilities may own their own generation, transmission, and distribution facilities and may also perform their own system operations. Smaller munis usually band together to create public power agencies that share ownership of generation and transmission. Virtually all munis own and operate their own distribution systems[2]. Many munis also buy power directly from federal power agencies and in recent years have commonly traded power with IOUs and other parties. The U.S. has over 2,000 munis and PUDs which serve about 14% of end-use customers.

Rural Electric Co-ops

As the electric grid evolved in the United States, neither municipal utilities nor IOUs had much interest in building costly distribution systems into rural areas. Given concerns about quality of life and the sustainability of agriculture, the federal government created the Rural Electrification Administration in 1936 which provided for the creation of rural electric co-ops. Co-ops are utilities owned by their customers (called members) and run by an elected board. Co-ops own all the distribution lines within their area and provide all electrical service to their customers. Many co-ops are distri-

---

[2]The exception is that in recent years some municipalities have created aggregation utilities. These do not own facilities, but buy power on behalf of their citizens and use the IOU's distribution system to deliver that power to customers.

bution-only utilities which purchase their power from federal generation agencies. In cases where there is not enough federal generation available, groups of co-ops have banded together regionally to create generation-and-transmission cooperatives that own facilities on behalf of the distribution co-ops. Co-ops are operated as not-for-profit organizations and any excess funds collected are returned to the members at the end of the year. The U.S. has just under 900 co-ops that serve about 12% of U.S. customers.

Federal Power Agencies

Federal power agencies are entities created by the U.S. government to

LARGEST MUNIS IN THE U.S.	
	1,000 MWh
1. Salt River Project	27,694
2. City of Los Angeles	24,317
3. City of San Antonio	18,892
4. Long Island Power Authority	18,751
5. City of Memphis	15,256
6. Jacksonville Electric Authority	12,844
7. Nashville Electric Service	12,831
8. South Carolina Public Service Authority	11,592
9. Austin Energy	11,547
10. Sacramento Municipal Utility District	10,818
2007 data from EIA	

market the power output of federal projects – primarily hydro power on federal dams. There are five federal power agencies – Bonneville Power Administration (BPA), Southwestern Power Administration (SWPA), Southeastern Power Administration (SEPA), Western Area Power Administration (WAPA), and the Tennessee Valley Authority (TVA)[3]. The federal power agencies generally perform the role of generation provider to a vertically-integrated public utility sector (i.e., to munis, PUDs and co-ops). Due to federal law, preference in the sale of federal power must be given to public bodies and co-ops. IOUs can buy federal power only if it is surplus power that cannot be sold to the preference customers. In some limited cases, the agencies are also authorized to sell directly to large industrial customers. The agencies also own transmission lines that run from their projects to other utility-owned grids. In some cases, such as in the western U.S., these transmission lines can be quite extensive. In these areas, it is common for the federal power agency to also assume the system operations function.

---

[3] The TVA is technically not a power agency, but is very closely related in function and legal authority. Because the TVA also has responsibilities related to economic development it has added significant additional generation such as nuclear power plants.

Public Power Agencies

As described above, smaller municipal utilities, PUDs and co-ops often work together to own generation and transmission. They do so by creating a public power agency. These agencies are owned by the participating utilities and take on the responsibility of owning and operating generation and transmission facilities. By joining together, the smaller utilities are able to share the costs and risks of owning and maintaining the facilities necessary to serve their customers.

Power Pools

Power pools are created by groups of utilities that turn over to them the scheduling and dispatch function for their power plants. The concept behind the power pool is that multiple utilities within a region can gain higher reliability and lower costs by placing their generation assets into a regional pool. The units can then be operated on a regional basis. This results in cost savings to all participating utilities since higher-cost utilities have access to generation that may be lower-priced than their own, and lower-cost utilities receive additional power sales revenues. By using a pool, utilities do not need to continually rely on trying to trade power to obtain the savings, but rather can obtain them through the routine scheduling and dispatch process. Power pools were used extensively in the northeastern U.S. prior to deregulation, but the largest of the pools – PJM, New England and New York – have now been replaced by ISOs.

Energy Services Companies (ESCOs)

Energy services companies, or ESCOs, evolved in the regulated model to offer services beyond the regulated services offered by utility companies. Typical ESCO services include bill evaluation, demand side management, appliance maintenance, power reliability, and power quality. These services are provided by numerous for-profit organizations, both large and small, and are in addition to regulated services offered by utilities.

Independent Power Producers and Electric Marketers

As competition came into vertically-integrated markets, limited roles for independent power producers (IPPs) and electric marketers evolved. Independent power producers are non-utility for-profit companies that own generation and sell the output under long-term contracts. Electric marketers are entities that buy excess supply from generators and/or utility companies and resell the power to other market participants. These roles are generally very limited in the vertical utility model and we will discuss them in more detail below. In a vertically-integrated market, the role of the IPP is to offer an alternative to utility-financed construction. Many IPPs build power plants and sell the output to the utility under long-term contracts. For example, a utility that needs 200 MW of new supply might find it advantageous to contract with an IPP rather than building the capacity itself.

Participants in Restructured or Competitive Electric Markets

As electric markets are restructured to allow competition in the generation or retail sales sides of the business, roles are opened for a significant number of new market players. We will explore different competitive market structures in Section Nine, but for the purposes of this discussion it is important to understand that restructured markets typically move generation, system operations and retail sales (meaning the sale of electric supply to end-use customers) outside the purview of the monopoly utility. The utility becomes a transmission and distribution (or sometimes just distribution) organization that delivers electricity to end-use consumers on behalf of other market participants.

Merchant Generators

Merchant generators are independent owners of generation that are not part of the regulated utility. They own and operate generation in the interests of making a profit for their shareholders. Unlike IPPs, who tend to contract all their capacity to utilities, merchant generators sell to a variety of market participants and are generally more exposed to market prices. Merchant generators offer a number of services such as electricity (MWh), capacity (MW) and/or other ancillary services which may be sold to utilities, marketers, ISOs, or directly to end-use customers. Merchant generators may build new generating stations or may acquire units from utilities selling off existing generation. Some merchant generators evolved as unregulated subsidiaries of utility companies while others formed as independent companies. Examples of merchant generators include Calpine, Constellation Generation, Exelon Generation, and Luminant.

Transmission Companies

Transmission companies, or transcos, are independent owners of transmission facilities. They are investor-owned, and like IOUs they operate to make a profit for their shareholders. Because they are monopolies they are regulated by FERC. They acquire their transmission lines either by buying them from formerly vertically-integrated utilities that have decided to divest of the transmission business or by building new transmission facilities. Some market analysts envision that over time transcos will evolve to become a combination of transmission owner and system operator, but that has not yet occurred in the United States. Examples of major transcos in the U.S. include American Transmission, ITCTransmission, National Grid USA, and Trans-Elect.

Independent System Operators (ISOs)/Regional Transmission Organizations (RTOs)

If competitive generation markets are to work effectively, generators must have non-discriminatory access to the transmission system to deliver their power to customers. The traditional structure in the U.S. is for transmission systems to be owned and operated by the utilities. This allows the utilities to control access to transmission since, under federal law, they are allowed to provide preference for service to native loads (their own customers). And because they also control the system operations function, they decide which units are dispatched in response to system needs.

Merchant generators and marketers argue that this creates an unfair playing field in the market. One solution is to create an Independent System Operator (ISO) or Regional Transmission Organization (RTO). An ISO/RTO does not own transmission, but rather manages transmission owned by other entities (either utilities or transcos). The ISO/RTO handles all the system operations functions of scheduling generation, transmission, and reserves; acquiring other ancillary services; and managing the system in real time. ISOs/RTOs are non-profit organizations run by an independent board of directors that is not beholden to any one market participant or group of market participants. ISOs/RTOs in the United States are FERC-regulated entities, with the exception of the ERCOT ISO in Texas[4]. ISOs/RTOs in North America include the Alberta Electric System Operator, California ISO, ERCOT ISO, ISO New England, Midwest ISO, New York ISO, Ontario Independent Electric System Operator, PJM Interconnection, and SPP RTO.

Electric Marketers

Marketers generally purchase electricity from generators and then resell it to utilities, end users or other marketers. Successful marketers add value by saving generators and customers the trouble of finding each other, arranging for transmission and ancillary services, and sometimes assuming price or other marketplace risks. A number of generation companies have an electric marketing arm to handle marketing of their units' output, while others simply sell to independent marketing companies.

---

[4] Due to the configuration of transmission lines in Texas (they generally don't cross state lines) ERCOT has successfully maintained its status as state-regulated.

The role of the marketer is sometimes divided into two categories – the wholesale marketer and the retail marketer. Wholesale marketers buy power and resell it to utilities, other marketers and very large industrial customers. Retail marketers also buy power, but focus solely on resale to end-use customers. Since retail marketers usually have a lot more customers than wholesale marketers, their skills are focused on mass sales, customer service, product development, billing, credit and collections, and brand development. Wholesale marketers tend to focus more on risk management and direct sales. Examples of active electric marketers include BP, Constellation NewEnergy, Direct Energy, Suez, Sempra, and TXU Energy.

Financial Services Companies

Financial services companies provide risk management services associated with price risk and other risks that are inherent in the electricity industry. Electricity prices can be extremely volatile, and many market participants cannot handle the cash flow impacts of rapidly changing prices. Thus there is a market need for entities that can offer hedging products to lay off price risk. We will discuss this function further in Section Fourteen. Financial services companies active in the electric industry include J.P. Morgan, Bank of America, Goldman Sachs, and Deutsche Bank.

Transmission Owners

In markets where the system operations function has been moved to an ISO, the term transmission owner, or TO, is used to describe the entity that continues to own, maintain and, if necessary, expand the transmission system. The TO operates the transmission under direction from the ISO and receives revenues from the ISO to cover the cost of ownership and expense of operation of the transmission line. TOs may be transcos or they may be part of a utility distribution company.

Utility Distribution Companies

As markets are restructured, many of the functions previously served by the monopoly utility are removed from the utility function and replaced by competitive companies. This can include generation, system operations, transmission, and retail sales of supply. The one remaining utility function is the distribution function. Thus in markets where generation and retail sales have been made competitive, the utility company becomes a UDC, or utility distribution company. The UDC is the monopoly provider of distribution services. This may also include providing supply to some customers – usually smaller customers who are not eligible for competitive services or larger cus-

SECTION EIGHT: MARKET PARTICIPANTS IN THE DELIVERY CHAIN

tomers who have chosen not to take service from a marketer. In other cases, regulation prohibits the UDC from offering supply services.

Load Serving Entities

In many regions, the term load serving entity (LSE) is used to refer to any market participant that provides supply to end-use customers. This may be a UDC or it may be a marketer. Other terms used include energy services provider (ESP) or retail electric company (REC).

Energy Services Companies (ESCOs)

In competitive markets, ESCOs continue to provide an important market function. In addition to providing the services described in the vertically-integrated market model, they may also assist larger end-use customers in evaluating the various market options and finding optimal solutions for acquiring supply and satisfying electric needs.

What you will learn:

- What an electric market structure is
- Goals of an electric market structure
- The electric market structures currently found in the U.S.
- How electric market structures function
- How different market structures address day-to-day system operations

9

9

SECTION NINE: ELECTRIC MARKET STRUCTURES

In the last few sections, we have learned about the necessary components of a functioning electric system. Generation units must be built to provide energy and reserves, transmission lines must be built to move large amounts of power over long distances, and distribution systems must be built to distribute electricity safely to end-use consumers. And, of course, power system operations must tie it all together by dispatching units to keep supply and demand in balance at all times.

The next issue we will address is how to structure a marketplace that allows all this to occur in a cost-effective and functioning manner. Capital must be attracted which will allow sufficient generation, transmission and distribution to be constructed. Markets must be designed which will allow the various sectors to interact and provide the various necessary reliability functions. And all the while the market structure must foster delivery of reliable and low-cost electricity to residential and business consumers.

Due to mixed efforts at market restructuring, we currently have varied electric market structures in place across the U.S. The story of how we got to where we are today – and where we may be going – will be told in Section Twelve. Here we will introduce four market structure models that are employed in the U.S. today, and then discuss how each model addresses the necessary functions of electric generation, ancillary services, transmission access, and balancing supply and demand in real time.

What is an Electric Market Structure?

An electric market structure is the set of rules and responsibilities that defines how market participants interact with each other to provide electricity to consumers. The key questions to defining a specific market structure are:

- Who is allowed to own generation units and who will buy their output?
- Who will schedule unit generation, reserves and transmission access in the forward market, and who will manage the system in real time?

SECTION NINE: ELECTRIC MARKET STRUCTURES

- Which customers, if any, will be allowed to buy supply directly from marketers or generators and which customers must buy from the distribution utility?

- How are business transactions performed to allow supply and system operations needs to be acquired by market participants?

The market structure that exists in any specific marketplace is determined by a combination of federal, state, or local legislation and/or federal and state regulatory decisions. Ultimately, an electric market structure should benefit consumers – specifically it should create reasonable prices, reliable service, fairly predictable bills, and it should encourage innovation in services. The definition of a reasonable price, of course, is open to significant debate, but in general it means high enough to keep the supplier in business and able to invest in necessary infrastructure, while low enough to be affordable enough that business consumers are not put at a disadvantage relative to other states and/or countries.

> **MARKET SECTORS THAT DEFINE MARKET STRUCTURE**
>
> **Generation** – The generation of electricity.
>
> **Transmission** – The movement of electricity at high voltages, usually between generation and distribution systems.
>
> **Wholesale Trading** – The trading of electricity between entities that are not end-use customers (such as generators, marketers and utilities).
>
> **System Operations** – The operation of the generation and transmission grid to ensure supply matches loads at all times and that system reliability is maintained.
>
> **Distribution** – The delivery of electricity (which may be owned by someone other than the distribution company) from the transmission system to the end-use customer.
>
> **Retail Sales** – The sale of electricity and other value-added services to the end-use customer.

A key component of any market structure is the existence of – or lack of – competition. In general, there are three areas where competition has the potential to benefit consumers – generation of electricity, wholesale trading of electricity and sales of electricity to retail customers. System operations, transmission and distribution are natural monopolies, at least in today's technological environment, and are not open to competition. We all pretty much agree that it does not make economic, environmental or aesthetic sense to build duplicative transmission or distribution lines. And the very nature of system operations necessitates it be performed as a single centralized function. So in discussing different models for electric markets, we will really only be discussing different levels of competition in generation, wholesale trading and retail markets. We will also discuss the players that own and operate the transmission, distribution, and system operations monopolies, and how these functions are restructured to support competition in the competitive sectors.

UNDERSTANDING TODAY'S ELECTRICITY BUSINESS

In this section we will discuss four basic market structures:

- Vertically-integrated monopoly utility
- Single buyer with competitive generation
- Wholesale/industrial competition
- Complete retail competition

You should remember that these are only models, and actual markets may vary from them in the details.

Vertically-Integrated Monopoly Utility Model

The vertically-integrated monopoly utility model arose as electric service began in the late 1800s and survived unquestioned for close to one hundred years. The concept behind this model is to treat generation, transmission, distribution, retail sales, and system operations functions as an integrated whole – owned and performed by one monopoly entity or by closely-aligned monopoly entities. In the United States, three different sub-models evolved within the vertically-integrated model:

- The investor-owned utility (IOU)
- The municipal utility and public utility district (Muni and PUD)
- The rural electric co-op (co-op)

The majority of U.S. customers are served by IOUs. Traditionally, IOUs have served their customers by running the generation, transmission and distribution systems as a vertically-integrated system designed to serve the needs of all end-use cus-

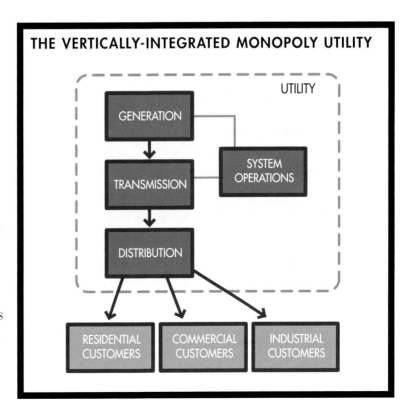

tomers within their service territory. Sometimes IOUs trade with other IOUs to minimize supply costs and/or enhance reliability. In other cases this is done through power pools. But in general, under vertical integration the IOU owns everything and serves everyone within its territory.

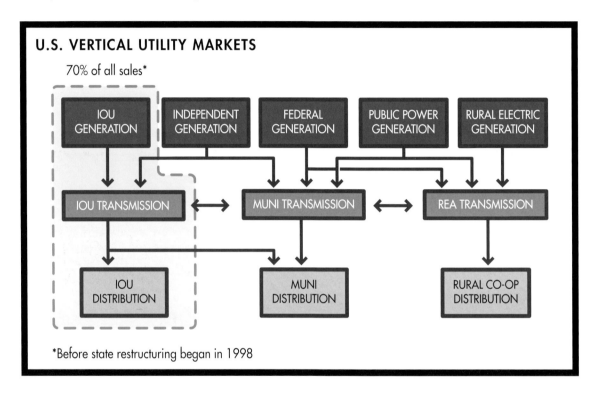

In other areas, munis or rural co-ops have evolved as the distribution utility serving all customers within the utility's service area. Large munis operate much like IOUs with complete ownership of generation, transmission and distribution. But smaller munis and co-ops do not have the necessary volume of customers to make it economical to own the complete vertical chain. These entities do own their distribution systems, but either band together to jointly own generation and transmission in public power agencies or depend on federal generation agencies to supply the power to them.

The Current Status of Vertically-Integrated Monopoly Utility

Many parts of the United States still operate under the vertically-integrated model. These include most of the Southeast, the Northwest and much of the West outside of California. Virtually all munis, PUDs and co-ops still operate under the vertically-integrated model with the exceptions noted above for sharing generation and transmission resources.

Single Buyer with Competitive Generation Model

The first movement towards introducing competition in an electric marketplace must come through the entry of competing electric generation providers. Without this, there can be no competition in other sectors. Thus, the second market model is one in which the utility maintains the monopoly functions of transmission, system operations, distribution, and retail sales, but in which non-utility generation is allowed to compete with utility generation. Initially in the U.S., only cogenerators were allowed to compete with utility generation, but in the 1990s changes in federal law allowed other non-utility generators to enter the marketplace.

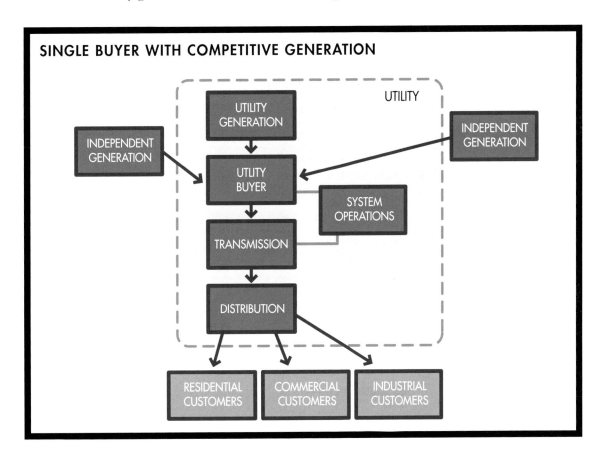

Under the single buyer model, the utility company creates a supply purchasing group whose job is to competitively procure supply on behalf of all the utility's end-use customers. This power may be procured from the units owned by the utility as part of its regulated function or from merchant units owned by competitive entities. Under this model, system operations may be handled either by the utility or an Independent System Operator. Some wholesale trading outside of the utility may occur, but since the utility is the only buyer in the local market this happens only if a marketer is aggregating supply from generators and then selling it to the utility, or if the marketer is moving power across the utility system into another market. The remainder of the market functions – transmission, distribution and retail sales – continue to be the monopoly purview of the regulated utility.

Generally, the utility supply purchasing group makes its buying decisions based on reliability of supply and least cost. Since utility generation already has its capital and debt costs covered in customer rates, it does not need to capture these fixed costs in its prices and can offer lower bids. Thus, this model tends to result in independent generation selling into the utility only when existing utility generation is not sufficient to cover customer loads. If the utility is an IOU, it is necessary for the state regulatory body to create mechanisms to review its purchasing decisions to ensure they are in the interests of ratepayers, and not the IOU's shareholders.

Many market participants will argue that the single buyer model is really very limited competition. While generating units may be owned by entities other than the utility, their only source of revenue is sales to the utility. Thus there isn't much in the way of true competition. To get true competition, markets must allow producers and consumers to come together outside of the regulated monopoly.

The Current Status of Single Buyer

Many states that otherwise subscribe to maintaining the utility monopolies have opened up their generation sectors to limited competition by implementing some form of the single buyer model. A common implementation of this model in recent years has been the requirement that utilities needing new supply sources must consider power contracts with independent parties in addition to utility-constructed units when doing integrated resource planning. In some states using this model, new generation has been constructed by merchant generators who then sign long-term (7-10 year) supply agreements with the utility. Other states have implemented periodic auctions where the utility buyer acquires supply from the market using a centralized auction process.

Wholesale/Industrial Competition Model

Many observers of electric markets believe that the real benefits of competition are to be gained by allowing only large customers – the largest commercial and industrial customers – to competitively procure supply while the smaller customers continue to be served by the utility company under regulated rules. Their argument is that, at least initially, only industrial customers really care enough to get involved in competitive procurement and are sophisticated enough to look out for their own well being. This allows the competitive market to mature before deciding whether there are benefits to providing supply choice to smaller customers.

This philosophy leads us to the wholesale/industrial competition model, in which large commercial and industrial customers purchase their electric supply directly from

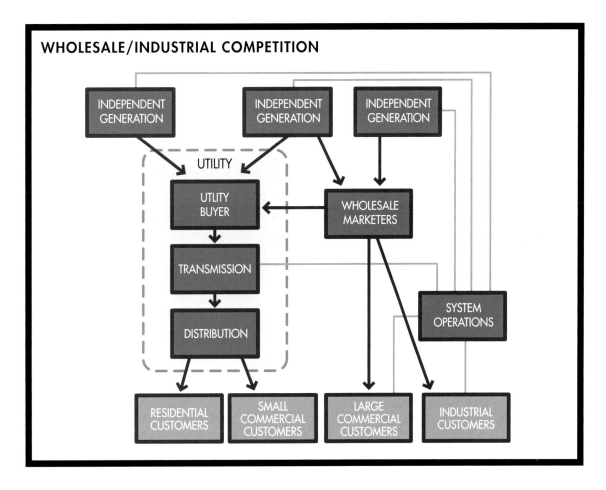

generators or wholesale marketers, but the utility maintains the single buyer function for small commercial and residential customers. For this model to work, there must be an independent entity responsible for system operations. Since the system operator implements the rules that determine who gets access to transmission and whose units are ramped up or down in response to system needs, leaving the function within the utility could unfairly bias the markets. Thus an Independent System Operator (ISO) is created to handle the system operations functions. This market structure also creates the need for entities to match generators with end users, so the role of the marketer becomes important.

Transmission continues to be owned by the utility, but is operated under the direction of the ISO[1]. The utility continues to own and operate the distribution system, providing distribution-only services to large customers and providing bundled distribution/supply services to smaller customers. The utility supply purchasing group acquires supply for all the smaller customers and, as in the previous model, has the option of obtaining supply from utility generation or through contracts with independent generators. In this model, however, a competitive wholesale market is likely to evolve. Because of this competition, smaller customers should see some benefits if it does result in lower wholesale prices since the utility buyer should have access to the lower-cost supplies.

Key issues for the regulator in this model include defining the arrangements the utility buyer is allowed to enter into to buy supply and determining how the cost of that supply is passed on to the customers who purchase it from the utility. Options for purchasing supply include bilateral contracts, spot purchases and periodic auctions. Options for passing on costs include monthly or annual pass-through of costs, or a rate cap that puts the utility at risk for costs above the cap.

The Current Status of Wholesale/Industrial Competition

Many of the states that have undergone electric deregulation have implemented wholesale/industrial competition or some variation thereof. A common variation is to give smaller customers the option of buying competitive supplies but to allow the utility to continue to provide default service (continued bundled distribution/supply service for customers that do not choose to acquire supply from a marketer). In most situations, smaller customers have overwhelmingly chosen to remain on default service, making the market a de-facto wholesale/industrial competition market. Thus this model best represents the majority of states that have implemented deregulation.

Complete Retail Competition Model

Under complete retail competition the journey to a competitive market is complete. In this model, the utility has been completely removed from the supply function on

---

[1] An alternative would be for independent transmission owners, or transcos, to take over the transmission systems. In theory, they could also take over the system operations functions and act much like interstate pipelines in today's natural gas marketplace. While we have seen some utilities sell off their transmission to regulated transmission companies, we have yet to see any examples of these companies also taking on the role of an ISO. This model is used in other countries such as Britain.

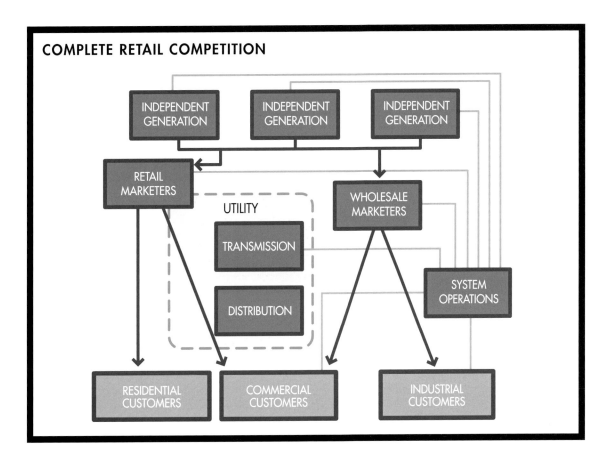

both the generation and the retail sales side. The utility is now simply a transporter of electricity that provides the transmission and/or distribution infrastructure used by various market participants on an open access basis. An ISO is required to perform the system operations functions in an unbiased manner. Wholesale and retail marketers acquire supply from generators and sell to each other as well as to end-use customers. Generators can sell directly to customers or to a marketer. Under this model, the regulators must create a mechanism for fulfilling the function of provider of last resort (POLR). The POLR serves customers who are unwilling or unable to contract with an electricity provider. This may be done by assigning customers to marketers on a pro-rata basis or by selecting a specific marketer to provide this service under regulated terms.

The Current Status of Complete Retail Competition

Although many states now allow choice of supplier in some form or another to all customers, most still allow customers to choose to continue taking supply from the utility company. So far, participation in customer choice by smaller customers has been minimal based on various market and regulatory impediments. Only a limited number of

SECTION NINE: ELECTRIC MARKET STRUCTURES

THE MARKET MODELS				
Function	Vertically-Integrated Monopoly Utility	Single Buyer with Competitive Generation	Wholesale/Industrial Competition	Complete Retail Competition
Providing generation	Utility	Utility and IPPs	Merchant generators	Merchant generators
Purchasing electricity	Utility	Utility	End users for large customers, utility for small customers	End users
Power plant scheduling and dispatch	Utility	Utility or ISO	ISO	ISO
Transmission scheduling	Utility	Utility or ISO	ISO	ISO
Running the forward spot market	Utility	Utility or ISO	ISO	ISO
Ensuring supply and demand are balanced in real time	Utility	Utility or ISO	ISO	ISO
Managing distribution functions	Utility	Utility	Utility	Utility
Providing supply to end users	Utility	Utility	Utility for small customers, marketers or generators for large customers	Marketers or generators
Long-term planning	Utility	Utility	Utility for small customers, market for large customers and ISO for both	Market and ISO

states have fully implemented the complete retail competition model, where utilities are completely removed from offering supply to end-use customers. Many other states have a mix of the complete retail and the wholesale/industrial model where both utilities and marketers may supply smaller customers.

Trading Arrangements

The four models discussed above answer all of the questions that define market structure with the exception of the last, which asks how business transactions are performed to allow supply and system operations needs to be acquired by the marketplace. As we move away from the vertically-integrated utility model, we begin to take the supply and system operations functions from the utilities and turn them over to the marketplace. Given the physical complexity of the electric marketplace, these functions cannot be left to chance. Regulators and legislators must carefully devise market-based arrangements to ensure that participants can efficiently acquire generation and that system operations functions continue to be provided. These rules are called trading arrangements.

> **TRADING ARRANGEMENTS**
>
> - Generation scheduling and dispatch
> - Ancillary services scheduling and dispatch
> - Transmission access
> - Managing imbalances in real time

Trading arrangements answer the following questions:

- How do buyers arrange for electric supply in forward markets?
- How do buyers arrange for electric supply in spot markets?
- How do buyers arrange for ancillary services?
- How do market participants receive access to the transmission system?

The three basic ways of structuring trading arrangements are wheeling, decentralized and integrated. Each is discussed in detail below.

Wheeling

Trading arrangements under wheeling are applicable only to the vertical utility and the single buyer models. Under the wheeling method, each utility schedules its own generation plus any purchased power in the day-ahead based on the utility's load forecast. Ancillary services are scheduled by the utility from its own generation or are acquired by the utility from nearby utilities or IPPs through bilateral agreements (a bilateral agreement is simply a private contract between two parties). Generation, purchased power and ancillary services are scheduled in an integrated manner to provide the lowest-cost service subject to transmission and other constraints.

Imbalances between scheduled generation and loads in real time are managed by the utility's system operator who simply ramps up or down utility-owned or contracted units that the operator has under full control. The transmission system is made available for use by parties who wish to wheel power (i.e. move the power across the utility transmission system for delivery to a neighboring utility) on an open access basis, but only to the extent that wheeling transactions do not impact the utility's service to its own customers (native load). Transmission services, and any necessary related ancillary services, are made available to the wheeling parties under regulated tariff rates. The wheeling model is currently used in areas where there is no ISO.

Decentralized

The decentralized method is one of two ways of structuring trading arrangements that are applicable to competitive markets with independent generators, an ISO and end-users able to buy directly from marketers. The decentralized model moves as far away from the concept of a centralized market as is possible. An ISO is still required to handle scheduling, acquisition of ancillary services, access to transmission, and to manage the system in real time. But in this model, the ISO is seen mostly as a scheduler and an arbiter of free markets. For electricity bought and sold for future periods (forward markets), all contracts are simply bilateral agreements between generators and end users or entities that supply end users (load serving entities or LSEs). To obtain access to transmission, entities wishing to move power (known as scheduling coordinators) submit balanced schedules to the ISO that match specific supply to specific loads for each hour. Upon receiving the schedules for a specific hour, the ISO runs its power flow model to determine whether all requested schedules are feasible. If not, transmission congestion exists and access to transmission is allocated by an auction methodology. Market participants bid for the rights to use certain paths (or offer how much they would need to be paid to change their schedules to relieve the congestion) and transmission access is granted to those willing to pay the most. The market clearing price (i.e., the lowest accepted price bid for access to the path) is paid by all users of the path in a transmission congestion charge.

Necessary ancillary services, in amounts that meet NERC criteria, are either self-supplied by each scheduling coordinator or are acquired from the market by the ISO on behalf of the customers. Customers that do not supply their own ancillary services pay their pro-rata share of ancillary services costs to the ISO in an ancillary services charge. The ISO acquires ancillary services through auctions for each service (i.e. spinning reserves, non-spinning reserves, etc.). Any generator in the marketplace that has not committed its generation in a bilateral contract can bid into the auction. The ISO stacks the bids and takes the lowest that will satisfy the needed capacity amount.

All successful bidders are then paid the market clearing price (i.e., the highest bid accepted in a specific auction). Lastly, the ISO creates a stack of units available to provide real-time balancing energy (units that are willing to either ramp up or down in real time). This stack comes from successful ancillary service bidders plus additional units with uncommitted capacity that offer energy into the real-time market. The ISO then manages the system in real time by following its bid stack to increase or decrease supply in response to demand. Scheduling coordinators who create an imbalance in real time (meaning they either have more or less generation than they do demand on the system among their generator/customer pool) are charged/paid an after-the-fact imbalance price that is based on the prices paid to the units ramped up or down in real time.

As you can probably see, the decentralized method gets very complex very quickly. For each hour of the day, ISOs are running numerous auctions for transmission congestion, ancillary services and real-time stacks. Market uncertainly abounds as costs for ancillary services, transmission access, and real-time energy change hourly, often in unpredictable ways. And costs for real-time imbalances are not even known until after the fact. The existence of complexity creates numerous opportunities for market participants to find market weaknesses that can be exploited. ISOs are completely dependent on the marketplace to provide necessary ancillary services and imbalance energy. If the supply in any market becomes constrained, prices can skyrocket and the ISO has no means of optimizing solutions by moving generation from one market to the other. Many observers believe that one of the market design issues that led to the California crisis was the use of the decentralized model. The multiple and complex auctions allowed marketers to "arbitrage" between markets and find opportunities for excess profits. In such a situation, it is difficult for an ISO to craft and implement all the necessary rules to prevent these types of activities. In the U.S., the two large ISOs that initially used the decentralized method (California and Texas) have recently transitioned or are currently in the process of transitioning to the integrated model.

Integrated

The integrated method is the second methodology that can be applied to competitive markets. This method attempts to strike a balance between the complexity of the decentralized method and the lack of competition in the wheeling method. It recognizes the need for centralized markets and generation dispatch, while creating ways for competition to be taken into account in the operation of centralized markets.

SECTION NINE: ELECTRIC MARKET STRUCTURES

FUNCTIONS WITHIN THE THREE TRADING ARRANGEMENTS MODELS			
Function	**Wheeling**	**Decentralized**	**Integrated**
Transmission scheduling	Done by utility as adjunct to scheduling for its own customers	Done by ISO, congestion management done by auction	Done by ISO as part of optimization model
Generation reserves scheduling (ancillary services)	Done by utility, wheeling customers are charged a regulated price for the service	Done by ISO in multiple auctions	Done by ISO as part of optimization model
Operating day-ahead spot market	Done by utility or power pool run by utilities	Done by ISO in some regions, in others it is handled in the bilateral market	Done by ISO as part of optimization model, may also be bilateral market outside of ISO optimization
Dispatching balancing units in real time	Done by utility	Done by ISO, units available determined by auction	Done by ISO, units available determined by optimization
Setting market price for day-ahead market	Limited market, price is set by bilateral wholesale transactions or by power pool	Done by ISO in some regions, in others this is set by bilateral transactions	Done by ISO, as part of optimization
Setting market price for balancing market	Managed internally by utility	Done by ISO based on bid price of last unit dispatched	Done by ISO, as part of optimization

Under the integrated method, the historical concept of a power pool is modified into the concept of an ISO that operates spot markets as well as other system operations functions. The ISO is responsible for scheduling units in the day-ahead, allocating transmission, scheduling reserves, and balancing supply and demand in real time. Rather than running each of these as a separate and distinct market, as is done in the decentralized method, the ISO operates all the markets in one integrated, optimized fashion.

Basically, the ISO uses the old system operations model from the vertical utility days which created optimized schedules subject to system constraints. But it replaces the model cost inputs (marginal cost in the utility days) with bids placed by market participants. Owners of units bid for what they wish to be paid to operate during a given hour and also what they wish to be paid to provide reserves. Suppliers of loads bid for what they are willing to pay to receive power during a given hour. Typically the ISO runs a day-ahead market that matches supply to demand given the price bids, then minimizes overall system costs subject to constraints by creating a system schedule. If there are constraints on the transmission system, the ISO determines which units will be dispatched at what levels based on their bids. Locational prices are created and users of constrained transmission paths pay a congestion charge based on the difference in prices between the two locations (this is called locational marginal pricing). In real time, the ISO manages imbalances through a real-time market that ramps units up or down based on bid price.

The Current Status of Market Structures and Trading Arrangements

As you have learned, the U.S. currently has highly fragmented electric markets. Variations of each of the market models and trading arrangements methodologies are used in various states. This fragmentation is due to the way electricity is regulated – i.e., responsibility is split between states and the federal government. This makes it very difficult to create uniform markets across the U.S. as different states have different goals and different ideas about how to reach them. The fragmented markets make it difficult for energy companies to do business across state lines, or move electricity from one region to the next and impede development of our electric infrastructure. Unlike the European marketplace, which is moving rapidly towards a common vision of customer choice and competitive markets, the U.S. is caught in a position of indecision between the desire to revert to vertically-integrated markets and the reality that in many regions the market has already evolved too far to go back.

What you will learn:

- Why the electric industry is regulated
- The goals of regulators
- The historical basis for regulation
- Who regulates what
- How regulators establish rates and rules
- The various types of regulatory proceedings
- What tariffs are
- The rate case process
- What incentive regulation is and how it works
- How regulation works for monopoly utilities
- How regulation works for wholesale sales

SECTION TEN: REGULATION IN THE ELECTRIC INDUSTRY

It is impossible to understand the electric marketplace without a comprehensive understanding of the role of regulation. Regulation exists to ensure that customers of electric utilities and other service providers are protected from a lack of competition. To protect the public interest, regulation defines the services that regulated market participants can offer, sets rates to be charged for those services, prescribes accounting systems to track costs associated with them, enforces safety standards, approves construction of major new projects built by regulated entities, and monitors market behaviors of industry participants.

As the electric marketplace evolves, traditional concepts of regulation are also evolving. This evolution has hardly been uniform. Some states have moved forward with significant restructuring of regulatory-defined market structures while others have refused to venture beyond the traditional models. Thus an understanding of regulation at both the national and local level, as well as an understanding of how regulation may be evolving, is critical to success in an electric marketplace that can be both highly regulated and highly competitive. In this section we will explore who the national and state regulators are and how they determine the rules and rates for the services they regulate.

Why Regulate the Electric Industry?

The electric industry must be regulated due to the existence of monopolies. A monopoly is a business situation in which a corporation – through market power or a government-granted franchise – is either the only company conducting business in a given industry or the sole source of a specific commodity or service. A "natural monopoly" occurs in an industry where characteristics of the industry tend to result in monopolies evolving. A good example is the electric utility industry, where a proportionately large capital investment is required to produce a single unit of output and where large operators can provide goods or services at a lower average cost than can small operators. Both of these conditions occurred in the electric industry in the early 1900s. Thus, what began as a competitive industry quickly evolved into a market with few competitors. While the concept of monopoly utilities was ultimately deemed beneficial to the public, the resulting extreme market power created the potential for excessive profits and unfair favoritism to certain customers. This in turn created the need for government oversight of electric services.

The relationship between regulators and utilities is often described as the "regulatory compact." This means that in return for government regulators granting exclusive service territories and setting rates in a manner that provides an opportunity for a reasonable return on investment, investor-owned public utilities submit their operations to full regulation. In this section we will discuss the history of regulation and then look at how current market restructuring requires modification of the traditional regulatory compact for certain electric industry sectors.

The Goals of Regulators

Regulators generally seek to:

- Minimize costs to consumers and provide relatively stable rates.
- Maintain a fair playing field by not allowing undue discrimination.
- Ensure reliable service.
- Maximize the efficiency of resource use.
- Minimize negative environmental impacts.
- Ensure safety.
- Encourage innovation in services to customers.

Any given regulatory body may choose to focus on some or all of these goals. It should be remembered that in the end regulators are political in nature, and their attention to specific goals is driven by the political realities at any given period in time.

The Historical Basis for Regulation

State Regulation

In the late 1800s, the utility industry developed in an environment of open competition. Most cities and states believed that competition between utilities kept prices down, and it was not uncommon to find cities with numerous utilities operating in open competition. In fact, competition became so fierce that price wars were common, often leading to the demise of all but one utility, which could then take advantage of the lack of competition by raising customers' rates exorbitantly! As the electricity market evolved it became clear that its capital-intensive nature resulted in market inefficiencies (too much money spent on duplicative facilities) and allowed well-financed companies to push less successful ones out of the market.

To address this issue, state and local governments saw two options: municipal ownership of utilities or regulation of those that remained owned by shareholders. Between 1896 and 1906, the number of municipal utility systems more than tripled. This led investors who had created new shareholder-owned utilities to realize that they were in risk of losing the market to government entities. In 1907 the largest utility association (the National Electric Light Association) joined with the National Civic Federation (a turn of the century big business advocacy group) in favor of state regulation of electric companies. Subsequently the states of Massachusetts, New York and Wisconsin created the first state regulatory agencies. By 1916, 33 states had created regulatory agencies to oversee electric utilities. By this time, the model of a vertically-integrated utility owning generation, transmission, and distribution facilities and providing service in a specific franchise area under state or local regulation was well entrenched.

Federal Regulation

On the federal level, regulation was emerging as well. In the industry's early years, Congress allowed utilities to build and operate dams without regulation. But in 1905, the federal government began to license dams and charge fees for their construction. In 1920, the Federal Water Power Act created the Federal Power Commission (the predecessor of today's FERC) whose role was to regulate rates, financing and services of companies licensed to operate dams.

By the late 1920s, the government began to be concerned with the emerging large utility holding companies. These holding companies owned multiple local utility operating companies through a complex web of subsidiaries. By 1929, seven utility holding companies controlled 60% of the power in the U.S. Although the local utilities were regulated, nobody was responsible for oversight of the holding companies. Concerns about the potential for market power and financial shenanigans became so widespread that the federal government felt compelled to act. It took initial action with construction of federal power projects which, rather than selling their output to IOUs, made the power available to public power agencies on a preference basis. This encouraged public power as competition to the IOUs.

SECTION TEN: REGULATION IN THE ELECTRIC INDUSTRY

Then in 1935 Congress enacted two new laws. The Public Utility Holding Company Act (PUHCA) required interstate utility holding companies to register with the Securities and Exchange Commission (SEC) and to follow specific rules. PUHCA also broke up holding company systems that were not contiguous and prevented utilities from investing in non-utility businesses. The Federal Power Act exerted federal power over electric sales across state lines and expanded the authority of the Federal Power Commission to include regulation of such sales. Thus large utilities had two tiers of regulation – operations for service to their own customers were regulated by state commissions while sales to neighboring utilities were subject to federal jurisdiction. Although the Federal Power Act referenced sales across state lines, future court decisions extended this jurisdiction to all sales between utilities using transmission lines, since physically once electricity is on the transmission line it is impossible to say whether it stays within a state or not[1]. PUHCA was highly effective in breaking apart holding companies – between 1935 and 1950, 759 utilities were separated from holding companies and between 1938 and 1958 the number of registered holding companies declined from 216 to 18[2].

Federal involvement in the utility industry was further strengthened by:

- The Atomic Energy Act in 1946 which established the Atomic Energy Agency (the forerunner of today's Nuclear Regulatory Commission) and gave it regulatory oversight of nuclear generation, and in later amendments required open access transmission by utilities licensing nuclear plants.

- The passage of major environmental laws beginning with the National Environmental Policy Act of 1969 and including the Clear Air Act of 1970.

- The Public Utilities Regulatory Policies Act of 1978 and the National Energy Policy Act of 1992 which ended the utility monopoly on generation and gave FERC authority to require utilities to allow third parties to move electricity across utility transmission lines.

- Various other FERC actions in the late 1990s to further its role in regulation of restructured competitive wholesale electric markets.

- The Energy Policy Act of 2005, which gave FERC jurisdiction over electric reliability and clarified FERC's authority to penalize participants that manipulate markets.

---

[1] Certain sales in Texas have operated under state regulation due to the fact that Texas has configured its transmission grid so that it can be argued that electricity cannot leave the portion of the state covered by the ERCOT ISO.

[2] PUHCA was repealed by the Energy Policy Act of 2005.

We will discuss many of the more recent federal actions in Section Twelve when we consider market restructuring.

Who Regulates What?

Regulation of the electric marketplace is split between federal, state and local jurisdictions. For vertically-integrated utilities, services including generation, transmission, and distribution on behalf of the utilities' own customers are exempted from federal jurisdiction. These activities are regulated by:

- The state commissions for IOUs.
- The local governmental entity for municipal utilities in most states (13 states regulate some aspects of municipal utilities including rates).
- The co-op board for rural electric co-ops in most states (19 states regulate some aspects of co-ops including rates).

In states where restructuring has broken up the vertical utility, the IOU's utility distribution function (the utility distribution company, or UDC) remains under the jurisdiction of the state commissions. In these states, most sales of electricity to end users by marketers are only lightly regulated, but often the states create a minimal set of rules that the marketers must abide by.

Once utilities begin selling power to other parties besides their own end-use customers, federal jurisdiction is applicable for those specific transactions (although state jurisdiction continues to apply to vertically-integrated activities associated with service to the utilities' own customers). Thus the FERC regulates power sales between utilities and other wholesale entities (other utilities and marketers) and transmission services not on behalf of a utility's end-use customers (commonly called wheeling services). This jurisdiction applies not only to wholesale sales and transmission service activities of IOUs, but of munis and co-ops as well[3]. FERC also regulates sales by any power plants and services by any transmission lines not operated as part of a vertical utility. So merchant power plants and transcos are subject to FERC, not state jurisdiction. This applies even to utility companies that own generation in subsidiaries separate from the UDC and sell the output to their own UDC. In short, if the service is not part of unified vertical utility, it is subject to FERC jurisdiction. This also applies to ISOs, even if they operate only in one state (with the exception of Texas).

---

[3] One exception is that sales of power by co-op generation and transmission companies to co-op distribution companies are not subject to FERC jurisdiction.

SECTION TEN: REGULATION IN THE ELECTRIC INDUSTRY

Power plant siting is subject to state jurisdiction, while most power plant environmental regulations are federally mandated. Many environmental regulations are enforced by the Environmental Protection Agency (EPA) while others are enforced by state agencies. Operation of nuclear power plants is federally regulated by the Nuclear Regulatory Commission.

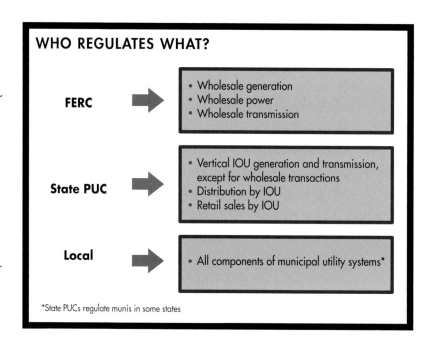

The Regulatory Process

Rates and rules are established through regulatory proceedings that are designed to give all interested parties a fair opportunity to state their opinions and present supporting facts. Regulatory proceedings include:

- Rulemakings — Proceedings held to establish new rules by which regulated entities conduct business.
- Rate cases — Proceedings that establish the rates a utility can charge for its services.
- Certificate cases — Proceedings that approve construction of new facilities.
- Complaint cases — Proceedings that evaluate complaints filed against utilities.

Following is a discussion of the general process used by regulators to set rates and rules for regulated services. Each regulator, however, may define this process in its own fashion.

The Initial Filing

A proceeding is typically initiated by a filing from a regulated entity (for rate cases and certificate cases) or by a market participant (for complaint cases). The documents filed are reviewed by the commission staff and a formal process is begun. Rulemakings are a bit different. They apply when a major restructuring or change in regulation is contemplated, and in this case the regulatory agency takes the lead. Thus they are initiated by the regulator, who prepares and publishes a proposed rulemaking that describes how the regulator suggests changing market rules and/or ratemaking.

Preliminary Procedures

Usually an Administrative Law Judge (ALJ) or one of the Commissioners is assigned responsibility for steering the case through the regulatory process. This person is charged with conducting the public hearings and with preparing a recommended decision for the full commission to consider. Prior to the start of the proceeding, a pre-hearing conference is often scheduled that allows any interested party to make an appearance and state the extent to which it will participate in the hearings. The party must identify the issues it will raise and is asked to state whether it will file briefs, submit evidence and/or cross-examine witnesses. These parties are then deemed intervenors in the case, and secure certain participatory rights in the proceeding. Following the pre-hearing conference, the ALJ or assigned Commissioner sets a date for hearings.

Hearings

Hearings are held to ensure the commission is aware of all important evidence relating to issues being considered. This is important because the commission must issue its decision solely based on the evidence presented in hearings (the evidentiary record). Prior to hearings, the intervenors generally file written documents (opening briefs) stating their position on the issues. Hearings are then held to provide evidence in support of the various parties' positions. Evidence may be entered through written documents (exhibits) or through written or oral testimony by witnesses. Witnesses are subject to cross-examination and all testimony is given under oath. Some bodies, such as FERC, depend mostly on paper hearings and rarely hold hearings with witnesses. In some cases, such hearings are replaced with technical conferences, where parties have an opportunity to state their positions but formal cross-examination is not used. At the conclusion of testimony, interested parties usually have another opportunity to file a written statement arguing their position (closing briefs), and then all parties have the opportunity to respond to each other's closing briefs (reply briefs). All closing and reply briefs are supposed to be based solely on the factual evidence presented in the hearings and no new evidence can be introduced at this point in the proceeding. Although this sounds like a very clearly defined process, it should be noted that commissions have wide latitude in how they run hearings and politics can often play a significant role in what occurs.

The Draft Decision

Following the hearings, a draft decision is issued by the ALJ or Commissioner, which is subsequently reviewed by the entire commission. Parties may file written comments on the draft decision for consideration by the full commission. Based on these comments, the ALJ or presiding Commissioner may revise his or her draft decision before submitting it to the full commission as the suggested action. The draft decision is by no means final, and represents only the opinion (educated, we hope) of the ALJ or assigned Commissioner. The commission as a whole decides on whether the draft decision or an alternative point of view will be final.

The Final Decision

After all comments have been filed, the full commission considers the draft decision at a hearing conference. Changes to the draft decision may be made by the Commissioners and occasionally two versions of a decision will be considered simultaneously. In this case, the decision different from the draft decision is called an "alternate decision." A decision becomes law when a majority of the Commissioners vote in support of it. At that point, it is called a final decision.

Review of Decisions

Any final decision is subject to review by the commission that issued it. Parties may request review either through a Petition for Modification or a Request for Rehearing. Petitions for Modification apply when a party believes that a decision fails to reflect the factual evidence presented in the evidentiary record. A Request for Rehearing applies when facts have changed since the evidentiary record was completed. If the commission denies review, or chooses to uphold the decision after review, the requesting party may ask the state (or U.S., depending on jurisdiction) courts to review the decision.

Tariffs

Tariffs are public documents, written by regulated entities and approved by the regulatory commission, that detail rates, rules, service territories, and terms of service. Tariffs are supposed to be written in accordance with the final decision of the regulatory body. Since decisions are often open to interpretation, tariffs must be approved by the regulatory body before they are legal. In general, tariffs include the following:

- A preliminary statement that describes the utility's terms of service and service territory and sets forth the accounts and adjustment mechanisms used in revenue accounting.
- Rate schedules that define rates and other terms of service for specific classes of customers.
- Rules that detail terms and conditions for service not described in rate schedules.
- Sample forms, including all standard form contracts, approved contract deviations and other standard forms used in day-to-day business.

Tariffs that have been approved by a regulatory commission are binding legal documents which constitute the contract between the regulated entity and its customers. A regulated entity cannot change its tariffs or fail to follow any provision in its tariffs in any way without approval from the regulatory agency. Copies of entities' tariffs can usually be found on the company's website.

Setting Rates through a Traditional Ratecase

A prime example of the regulatory process is ratemaking. One of the most important functions of the regulator is to set rates for monopoly services. The general concept of ratemaking is that monopoly entities are entitled to charge

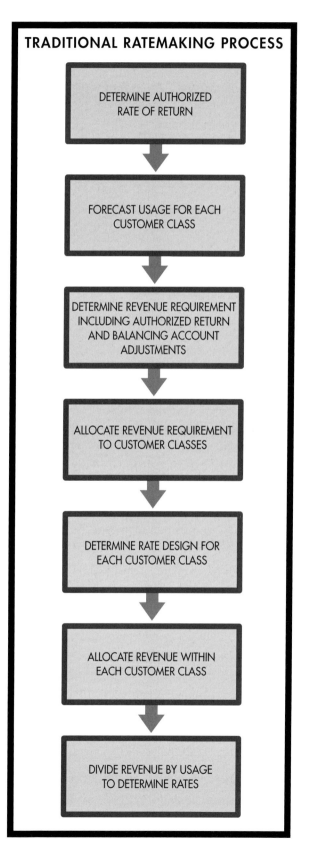

TRADITIONAL RATEMAKING PROCESS

- DETERMINE AUTHORIZED RATE OF RETURN
- FORECAST USAGE FOR EACH CUSTOMER CLASS
- DETERMINE REVENUE REQUIREMENT INCLUDING AUTHORIZED RETURN AND BALANCING ACCOUNT ADJUSTMENTS
- ALLOCATE REVENUE REQUIREMENT TO CUSTOMER CLASSES
- DETERMINE RATE DESIGN FOR EACH CUSTOMER CLASS
- ALLOCATE REVENUE WITHIN EACH CUSTOMER CLASS
- DIVIDE REVENUE BY USAGE TO DETERMINE RATES

rates that will allow them to cover their costs of service, plus a reasonable rate of return (or profit) on capital invested by shareholders to build the necessary facilities to provide the service. The process of setting rates requires determining a revenue requirement that includes all the revenue the utility needs to collect to cover costs and make a reasonable return, and then translating that revenue requirement into specific rates for specific customers. This process is outlined below.

Determining the Authorized Rate of Return

The first step in the ratemaking process is a determination of the utility's authorized rate of return. This is set by the regulatory commission, sometimes as part of the rate case and other times in a separate proceeding called a cost of capital proceeding. The regulators look at the current investment marketplace and determine how much return investors must be offered to ensure they invest in utility stocks as opposed to other investment opportunities. Separate rates of return will be set for utility debt and for equity, which is the money invested by stockholders.

Forecasting Usage

The second step is to forecast how much electricity customers will need over the rate case period. This information is important because it will determine the amount of capital and expense dollars that will be required to provide reliable service and the revenue that the utility will collect. Forecasts are made using historical usage data, expected growth or decline in population and business activities and other societal trends. The forecast will be broken down by customer class so that costs and revenues can be determined on a per class basis.

Determining a Revenue Requirement

A revenue requirement is defined as the total amount of money a utility must collect from customers to pay all operating and capital costs, including its return on investment. A utility's revenue requirement is determined by forecasting expenses (operating and maintenance, administrative and generation, fuel costs for power generation, and taxes other than income taxes), depreciation, and income taxes for a rate cycle, and then adding to that the return on rate base plus any amounts (positive or negative) outstanding in the balancing account. The rate base is the depreciated value of

> **DETERMINING A REVENUE REQUIREMENT**
>
> Expenses
> +
> Depreciation
> +
> Income Taxes
> +
> Rate Base x Authorized Rate of Return
> +/−
> Balancing Account Adjustment

all the capital facilities the utility has constructed in order to provide services to its customers. The return on equity multiplied by the rate base multiplied by the percentage financed through shareholder investment is the primary component of profit for a monopoly utility.

A balancing account is an accounting mechanism that keeps track of the difference between the revenue requirement and the actual costs incurred or revenues obtained by the utility. Any differences covered by the balancing account are added to or subtracted from future revenue requirements, thus insulating the utility and its customers from risks of revenue deviations. Typical portions of the revenue requirement covered by balancing accounts include expenses such as fuel and revenue fluctuations due to energy use that differs from the forecast for utilities with revenue decoupling (see page 177).

Allocating Revenue to Customer Classes

Once an overall revenue requirement for a utility service is established, it must then be determined what portion will be paid by each class of customer. This process is called revenue allocation. Various allocation methods are used in different situations. The most simple method (equal cents per kWh) allocates costs based on usage. While simple, this approach is not necessarily an accurate way of assigning costs. Since many of the costs of an electric system are fixed, actual costs caused by customers are more likely to be based on the maximum demand that a customer puts on the system, and not on the amount of kWh used. Thus a more common – though more complex – method is to allocate costs based on the estimated cost of service to each customer category (cost-of-service). This allocation can take into account demand-based costs as well as usage-based costs. An even more complex method (equal proportionate marginal costs or EPMC) allocates costs based on the marginal cost of serving each customer category. The marginal cost methodology looks at the cost of serving one additional increment in each class, rather than using the average cost as is done in the cost-of-service methodology. Actual determination of revenue allocation can be complex and is commonly one of the most highly contested issues in regulatory proceedings.

Determining Rate Design

Once a revenue requirement has been determined and allocated to the various customer classes, the rates that each customer class will pay are determined in the rate design phase of the proceeding. But before actual rates can be set, the rate structure must be determined. Rates are structured in any number of ways, but typically they are divided into three distinct components:

- Customer charges — A per-customer charge independent of usage.
- Demand charges — These charges are based on the maximum demand (kW) placed on the system during a specified period of time and are usually applied only to larger customers (due to the prohibitive cost of metering demand for small customers).
- Usage charges — These charges depend on actual usage (kWh) during a period of time. For many customers the rates used to calculate usage charges stay the same no matter when the energy is used. For larger customers, however, it is common to charge a different rate for different periods of the day. This methodology is called time-of-use rates.

During the rate design process, the method of allocating revenue between the different types of rate charges is also determined.

Allocating Revenue to Charge Types

Once the rate structure is set, another allocation must occur. This is the allocation within a customer class that determines how much revenue will be applied to customer charges, demand charges and usage charges. Once this has been done, we now know how much revenue the utility is expected to collect from each charge type within each customer class.

Determining the Rate

Finally, the rates for each customer type are calculated by dividing the allocated revenue by the appropriate forecasted factor. For instance, a residential customer class that is allocated $1 million per month to customer charges and has 100,000 customers forecasted, would have a monthly customer charge of $1 million divided by 100,000 customers, which equals $10 per customer per month.

Incentive Regulation

The process presented above is often called cost-of-service ratemaking. One complaint about this method is that in the absence of competition, utilities may not have incentive to keep costs down. The only incentive is the fear that the regulator may disallow expenses that are deemed excessive. When regulators do disallow expenses, utilities feel unfairly penalized (since it is always easy to second guess a cost after the fact). This conundrum has led some regulators to move to incentive regulation. Incentive regulation is designed to avoid after-the-fact reasonableness reviews while aligning shareholder and ratepayer interests by allowing both sides to benefit from successful performance by the utility. Examples of incentive regulation include performance-based, benchmarking and rate caps, each of which is described in more detail below.

Performance-based

Performance-based regulation compares the utility's performance to a specific index. An example is the procurement of gas supply for power plants. The utility's cost of buying gas would be compared with a market index for gas prices in the utility's area. If the utility's cost is lower than the market index, the utility's shareholders and ratepayers split the savings. Conversely, if the utility's cost of gas is higher, shareholders and ratepayers split the increased cost. This, of course, provides strong incentive for the utility to pay close attention to its gas purchasing strategies. Other examples of performance-based regulation might include incentives/penalties for achieving a certain level of customer satisfaction or maintaining a certain level of service without any outages.

Benchmarking

Benchmarking regulation sets rates in the first year of a rate cycle using traditional methods. For future years, rates are set by a formula that increases them based on an appropriate inflation index and then reduces them based on a regulatory-determined productivity factor that the utility is expected to achieve. This is often called x-y regulation where the factor x represents inflationary increases and factor y represents productivity-based decreases. This encourages the utility to go out and find ways to enhance productivity and to keep costs below inflation since any difference between actual costs and revenues collected in rates is a 100% cost/benefit to shareholders.

Rate Caps

Under rate cap methodology, fixed rates are set by the regulator for a period of years. Any variation between actual costs and revenue collected under the capped rates is a 100% cost/benefit to shareholders.

Service Standards

To ensure that utilities do not let service quality decline in the interest of achieving incentive revenue, regulators often create specific measurable service standards. Failure to achieve these standards results in shareholder penalties.

Market-based Rates

In cases where market forces are strong enough to prevent potential monopoly abuses, regulators have allowed regulated entities to charge market-based rates. This is common in the wholesale marketplace where generators and marketers that do not have a large enough market share to control markets are allowed to charge market-based rates. This type of regulation is the fundamental principle that underlies competitive electric markets created by market restructuring. To be authorized to charge market-based rates, generators or marketers selling power in the wholesale market must receive a certificate from FERC that authorizes them to do so and must demonstrate a lack of market power to obtain such authorization. FERC retains the right to implement rate caps and/or penalties at any point should these entities be observed exerting market power.

State and Federal Rate Methodologies

As noted above, the states are responsible for regulating the activities of vertically-integrated investor-owned utilities as well as investor-owned utility distribution companies. The state commissions are generally referred to as the Public Utility Commission or Public Service Commission. States regulate the services that utilities offer, determine the rates they can charge for these services, regulate reliability and safety standards, and set standards for financial practices of the utilities. Virtually all states still utilize the traditional rate case methodology for most IOU and UDC services, although they have begun experimenting with incentive ratemaking in some cases. It is important to note that no two state agencies regulate exactly alike. So with 50 states, we also have 50 different ways of doing business.

FERC is responsible for regulating all wholesale transactions, transmission services and ISOs. Most wholesale transactions are regulated under market-based rates although entities that hold large market shares are required to use cost-of-service-based pricing. Transmission services and ISOs are regulated under a traditional cost-of-service-based methodology.

The Future of Regulation

As we will see in the next section, regulation of the electric industry has changed significantly over recent years and is still in flux. Because regulatory authority is shared between federal and state agencies, the U.S. electric industry is subject to a wide vari-

UNDERSTANDING TODAY'S ELECTRICITY BUSINESS

ety of regulatory methods and market structures. It is likely that electric regulation will continue to evolve well into the future.

What you will learn:

- What restructuring is

- The arguments for why restructuring of the electric industry can provide benefits to consumers

- How markets mature as they are restructured

- The necessary components for a competitive marketplace

- How these components are implemented

SECTION ELEVEN: THE CONCEPTS OF MARKET RESTRUCTURING

The term deregulation is heard commonly in discussions about the electricity market. This, of course, is a misnomer. There is no such thing as deregulation in a market where key sectors are and will remain dominated by regulated monopolies. Even in sectors such as generation and retail sales, where competition can work effectively, the nature of electricity will result in all market participants competing in an environment of ongoing regulatory involvement. Consider, for instance, that a system operator must keep supply and demand in balance at all times, that one market participant's actions potentially impact everyone on the grid, and that disturbances travel quickly throughout regions. And then consider the value our society places on electric service. You can easily see why regulation must play a role in all sectors of the business to ensure safe and reliable service. So again, deregulation is not really what we are experiencing in the electric industry – what we have is market restructuring. And by market restructuring, we mean changes in regulatory rules that alter control, ownership or regulatory mechanisms of specific industry sectors resulting in increased competition.

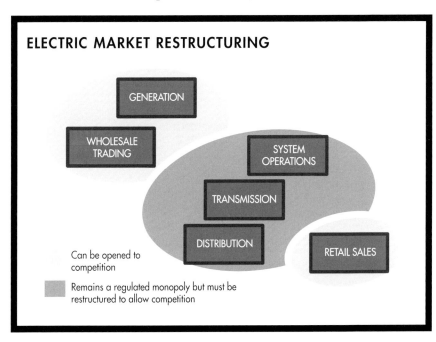

Due to the integrated nature of the traditional electric business, we must consider restructuring of all market sectors if we are to introduce competition into any of them. Failure to do so would create a facade of competition without truly creating an environment that can allow it to work. The sectors where we can introduce

127

competition are generation, wholesale trading and retail sales. Transmission, system operations, and distribution services are clearly a monopoly function, at least for the foreseeable future, but must be modified to allow for competition in other sectors.

We discussed what the different market structures might look like after restructuring in Section Nine. In this section we will discuss the concepts that support creating a new market structure and the implementation steps necessary to get there. These concepts are especially important to understand now because virtually all of the U.S. is in flux relative to market restructuring. Ten years into the restructuring process, the country is still stuck with no uniform vision. Some states and regions have moved far down the road to competition, others not at all. And unfortunately there is no clear federal policy for moving to unified markets. Thus, we can assume that anyone involved in the industry will be enmeshed in market restructuring uncertainties for quite some time.

Why Restructuring?

As you will soon learn, restructuring is a messy business. It costs a lot of money, market participants are forced to learn new ways of doing business, regulators lose control over market activities, and in some cases outcomes do not always benefit customers. So why bother?

The goal of a market structure should be to benefit customers by fostering reliable service, reasonable prices, fairly predictable bills, and to encourage innovation in the services provided. Many would argue that the first century of vertically-integrated utilities did an effective job of meeting those criteria, at least until issues with nuclear generation arose. Beginning in the 1970s, a number of utilities were forced to raise rates due to cost overruns on nuclear construction (much of which were due to changing safety regulations during the construction process) and the rise in fuel costs. Coupled with a general philosophical trend towards deregulation of U.S. industries (including airlines, natural gas and telecom) this led to a reconsideration of the benefits and necessity of vertical monopoly utilities. As the value of restructuring was (and continues to be) considered, a number of important questions arose:

- Can a competitive generation sector provide lower prices than regulated utility generation?

- Can competitive generation and wholesale marketing sectors result in adequate or improved reliability of supply?

- Will efficiencies resulting from a competitive wholesale trading marketplace result in lower prices?
- Will innovation in competitive retail services provide customer benefits that outweigh any negative aspects of competition?

Various market observers would answer these questions differently, and given the immaturity of electric deregulation, no one yet has the answers. But certainly deregulation has worked well in the U.S. natural gas markets, where prices have dropped (until the last couple of years), reliability has been enhanced and new services have been introduced.

The European Union, which in 1999 committed itself to competitive generation and retail sales markets across all of its member countries, lists its reasons for restructuring as follows[1]:

- To increase efficiency by introducing competitive forces into the electricity market.
- To eliminate distortions in competitive conditions that cause enormous price differentials between member states.
- To lower prices relative to the U.S. and Australia.
- To improve essential public services to all customers.
- To reduce needs for expensive reserve capacity by integrating markets.
- To reduce resource waste which results in pollution.
- To give customers the right to choose services that match their needs.
- To improve customer service provided by electricity companies.

Whether these benefits will be achieved is an open question, but few can argue the validity of these reasons for restructuring electric markets. In the end, it simply comes down to the question of what creates more benefits for consumers – a marketplace driven by competition or one controlled by regulation?

Market Evolution under Deregulation

Market restructuring is not an overnight process. It takes a long time to implement, there are lots of bumps in the road and benefits may not appear for as long as a

---

[1] Summarized from *Opening Up to Choice*, available at:
http://europa.eu.int/comm/energy/electricity/publications/doc/electricity_brochure_en.pdf

decade. Consider what has happened in the airline and telecom industries over the last thirty years. While the process has taken decades to evolve (and is still evolving) for these industries, does anyone doubt the benefits to consumers? For anyone old enough to remember the days of the vertical AT&T phone monopoly, simply consider the multitude of services available today compared to the plain black dial phones of the past.

To put the process of market evolution under deregulation in perspective, let's consider four phases of market evolution – regulation, deregulation, commoditization, and value-added services – as they provide an excellent framework from which to review the changes which may occur with restructuring in a specific region (remembering that different sectors will be at different phases in different states and regions in the U.S.). While the generation, wholesale trading and retail sales sectors will likely mature through these phases, they may not do so simultaneously, but rather will do so at different times in different regions. And in some regions they may even remain regulated or may go back to regulated after initially being deregulated. In almost all cases the transmission, system operations and distribution sectors will remain in the regulation phase, though regulation will have to be restructured to recognize the changes in the competitive sectors.

MARKET EVOLUTION UNDER DEREGULATION

Regulation
↓
Deregulation
↓
Commoditization
↓
Value-added services

Regulation

This phase is characterized by the dominance of regulation and lack of competition across the delivery chain. Transactions are generally highly structured and usually long-term in nature. Prices are fixed, buyers and sellers are relatively few, barriers to market entry are significant, vertically-integrated utilities dominate the marketplace, and customer choices are minimal. Prices are cost-of-service-based with little or no flexibility, and decisions to invest in infrastructure or innovation are highly influenced by support (or lack thereof) of regulators.

Prior to 1992, all of the U.S. electricity industry was in the regulation phase. Utilities or closely aligned generation agencies owned all generation, transmission, and distribution and operated their systems as a unified whole. Customers had little choice but to buy electric supply from their local utility.

Deregulation

In the deregulation phase, rules are loosened in some sectors and barriers to entry are broken down to allow competition to come into the market. As the number of competitors increases, transactions become more flexible and customers attempt to benefit from increasing choice and competition. Regulation still controls much of the way business is transacted, but is designed to encourage a level playing field among competitors and to foster competition in sectors that have been opened. Services in the competitive sectors (which may include generation, wholesale trading and retail sales) become more diverse and may be tailored to individual customers. While system operations remains highly regulated, prices for services such as reserves and transmission rights become market-based. Transmission and distribution prices remain cost-of-service-based, but incentive ratemaking is often adopted to encourage efficiency.

EXAMPLES OF ELECTRIC SERVICES IN THE MARKET PHASES

Regulation

- Utility distribution services
- Transmission services
- System operations in states without ISOs
- All sectors in states where full vertical integration remains in effect

Deregulation

- Electric supply to customers in states where the utility sells regulated commodity in competition with marketing companies
- Power purchases by utilities from merchant power generators
- System operations in states with ISOs

Commoditization

- Electric supply to industrial customers in states with robust customer choice programs
- Wholesale electricity trading in ISO markets and other regions where liquid trading exists (Western U.S. and Cinergy Hub)

Value-added Services

- ESCO services to industrial and commercial customers
- Financial services
- Combined commodity services (electric, gas, telecom, etc.) sold by a retail marketer

SECTION ELEVEN: THE CONCEPTS OF MARKET RESTRUCTURING

Commoditization

In the commoditization phase of the market maturation cycle, competition has taken hold. Numerous market participants compete with each other and trading volumes are high. Prices are market-sensitive and volatile. Regulations act mainly to prevent market manipulation and to assure fair access to monopoly infrastructure. Transactions become simplified and transferable among buyers and sellers. Financial markets arise where risks can be managed. Transactions that used to be secured with a phone call between old friends are now handled electronically with buyers and sellers often blind to each other's identity. The return on investment in infrastructure such as generation is purely based on market demand. Now shareholders of competitive companies carry the risk of bad management decisions. If a power plant is built and market conditions don't support its cost (as has happened to numerous such investments in recent years) no one pays except the shareholders (and maybe the bond holders) of the merchant generation company. Under commoditization, there is price transparency – meaning market prices are known to all participants – there are no barriers to transfer of commodity between willing buyers and sellers, no one entity has market power, and there are no regulatory protections.

Value-added Services

In this final phase, participants attempt to add value (and increase profits) by adding services their customers will value to the sale of commodity. In many instances, market

> **VALUE-ADDED SERVICES**
>
> **Facilities management** — Monitoring and maintaining a customer's equipment to increase efficiency of use and to minimize repair costs.
>
> **Energy management** — Monitoring customers' usage patterns and identifying ways to reduce energy costs through technological or behavioral modifications.
>
> **Demand side management** — Managing energy usage through the use of technology to reduce overall energy costs.
>
> **Pricing and risk management** — Offering pricing options to match energy pricing to the customer's business needs and appetite for price risk.
>
> **Power quality** — Using technology installed on the customer's premises to reduce voltage spikes and other power fluctuations.
>
> **Reliability** — Using technology installed on the customer's premises to reduce the probability of a power outage.
>
> **Combined commodity** — Combining electricity with other utility services such as natural gas, telecom and broadband to minimize customer hassle.
>
> **Billing options** — Offering alternate ways to pay for electric services and providing accounting and other related services.
>
> **Value-based** — Offering services based on value desired by the customer – hot water, conditioned air, lighting, etc. – rather than based on energy commodity.

restructuring has led to razor thin commodity margins, so marketers are forced to develop customer-focused services that will improve profits to the seller. Because one kWh of electricity delivered through the distribution system is the same as another, value-added services are the best way for participants to increase market share. In a retail electricity marketplace, most marketers must rely on value-added services to attain reasonable market share and profits. Low price alone is not enough to sustain a market position.

The Necessary Components for a Competitive Marketplace

The electric industry is different from any other industry for the various physical reasons we have discussed. And because of the complexities involved with operating the electric grid, it is extremely difficult to design a competitive market that functions effectively. The numerous vested interests that have built up over the past 100 years make it even harder to change anything. When changes are achieved, these same vested interests attempt to skew the new markets in their own favor (there is a theory that the California crisis was the result of a flawed market design purposely supported by participants that knew they could profit from it). The goal of a competitive marketplace is to create a liquid market with many buyers and sellers coming together in freely-negotiated business arrangements. Given the many peculiarities of the electricity business, it is necessary to clearly identify the components required for a functioning market and the necessary steps to implementing changes to achieve this market. First we will consider the necessary components, then we will discuss implementation.

Supply Side Competition

The first key requirement for a functioning competitive market is many sellers. In a perfectly competitive market no seller has enough market share to affect the market price of the commodity sold. With adequate competition, if any seller attempts to raise prices above what the market supports, customers can simply go to another supplier. Evaluating supply side competition in the electric industry is somewhat more complicated. In a given marketplace there may be many sellers for most hours out of the year, but during a few specific hours (peak hours or times when units are down for maintenance) there may no longer be a liquid market. During these times, certain suppliers may have the opportunity to exert market power. Worse yet, suppliers may be able to cause these situations to occur by purposely removing generation from the market (by taking it down for maintenance during a critical hour). Thus it is critical to encourage new generation construction in the marketplace, to encourage adequate transmission to let geographically remote suppliers compete, to find mechanisms that encourage utilities to divest of rate base generation, and to carefully review any regulatory protections that would favor existing generation units. In the absence of a fully competitive supply market, regulators must revert to regulatory solutions that include price caps, bidding restrictions, generation availability requirements, and/or profit controls. It should be noted that excess generation capacity is an additional requirement for a competitive market. There may be any number of generators supplying a given market, but without excess capacity the market is beholden to them in the same way it is if there is no competition.

Fair Access to Transmission

The need for many suppliers goes hand in hand with the need for fair access to transmission. If a supplier cannot get access to transmission, it cannot get its supply to market. Some markets in the U.S. have benefited less from new merchant generation than expected because transmission issues have prevented these units from fully functioning in the marketplace.

Establishing fair access to transmission is one reason for creating an Independent System Operator which, in addition to other responsibilities, is responsible for allocating transmission access. The ISO is not controlled by any market participant (unlike the wheeling model, where transmission access is controlled by the incumbent utility). Rules must be developed that give every market participant the same opportunity to use transmission and to allocate transmission access based on fair criteria – usually who is willing to pay the most. Any existing firm transmission rights contracts would ideally be bought out, otherwise certain market participants have favored access to the system. Rules must also be developed that determine how new transmission is paid for when it is required due to construction of new generation. If the costs are borne solely by the generators, new supply will be discouraged.

Unbiased System Operations

In addition to allocating transmission, the system operator must operate the system to ensure reliability and balance supply and demand in real time. This can work only if the system operator is in control of the generation assets serving the system. As loads grow, units must be ramped up. As loads fall, units must be ramped down. And sometimes, units must be redispatched from desired schedules due to locational issues. In a competitive market, generation owners are understandably reluctant to turn dispatch control of their units over to another entity unless they are convinced that the entity is running those units in an unbiased manner. Thus the requirement for a truly independent ISO.

Demand Side Competition

To ensure market liquidity, in addition to many sellers there must also be many buyers. One without the other does not make for a competitive market. Sellers cannot be tied to selling to a single utility purchasing entity – doing so simply creates a buyers' monopoly which must be regulated. To have a competitive market there must be the opportunity for end-use customers to choose to buy directly from suppliers rather than through their utility. Clearly the way to create the largest number of buyers is to completely remove the utility from offering supply services. This has been done and can work. But many states are reluctant to open markets simultaneously to all customers, and experience certainly has shown that there can be advantages to phasing-in customer choice. Many argue that a sufficient number of buyers can be created by allowing only large commercial and industrial customers the choice of supply. Since they make up significant demand in terms of load (though not in terms of number of accounts) this is often enough to create market liquidity. Another way to increase buyers is to also allow aggregated groups of smaller customers to buy directly. Some states are providing this opportunity through programs like municipal aggregation, where cities buy power on behalf of groups of customers.

A second key requirement on the demand side is to create markets where buyers become responsive to short-term electricity prices. By this we mean a market in which high wholesale prices will encourage buyers to curtail their usage until prices fall. Many current market structures do not provide for this as end users often pay average utility prices that are fixed for at least a year at a time. If wholesale prices spike to the equivalent of $5/kWh, end users in a traditional market have little incentive to care – they will likely continue to run their air conditioners at full tilt because they are paying the utility average price of $.10/kWh. This perhaps is one of the biggest factors in arguing for electric restructuring. As long as customers don't get price signals (and therefore do not have the opportunity to refuse the use of high-priced power), there is nothing to restrain market power and bad business decisions by utility executives and/or regulators except after-the-fact corrections by regulators. The key to providing for demand response to price is to create an infrastructure that gives customers the tools to do so. This requires hourly meters with data that can be accessed by customers in real time, some means of communicating hourly prices to customers, and an ISO market structure that allows loads to bid into markets in the same way that generation participates.

Distribution without Impediments to Competition

Also critical for a functioning competitive market are distribution services that foster, not impede, customer choice. There are a number of service issues that are critical:

- Providing access to meter data for both customers and marketers.

- Providing default suppliers for customers that don't choose or can't find a marketer.

- Determining whether the utility will be allowed to offer supply services, and if so setting an equitable price.

- Determining how societal programs such as demand side management and environmental programs will be paid for.

- Setting rules for customers that choose to self-generate, including stand-by rates and interconnection charges.

- Determining how transition costs associated with restructuring will be allocated.

- Determining how costs are allocated between distribution and supply services.

Many a budding competitive market has been stymied by just two of these factors — how the default rate is set for utility supply services (if it's too low, no marketer can compete) and how meter data is made available to marketers (without meter data, marketers can't bill their customers).

Opportunities for Hedging Risks

Prices in competitive markets — especially for a commodity like electricity — are volatile. Electricity prices fluctuate rapidly based on supply and demand. Demands at 6 p.m. are often double what they are at 6 a.m. This means twice as much generation is required — at much higher prices. And as gas prices fluctuate, availability of hydro power comes and goes, hot or cold weather drives demand, and large units go down for maintenance, price fluctuations from $20/MWh to $250/MWh are not unexpected. Many market participants cannot handle such fluctuations. Thus regulators must always be cognizant of creating stable rules and market structures that can help foster the development of financial markets for risk hedging.

Creating a Competitive Market

Now that you understand the requirements for a competitive electric marketplace, let's take a look at the issues involved in its actual implementation. As we have seen repeatedly in the U.S., the move from a regulated monopoly market structure to a competitive one is no simple task. Regulators and legislators have a number of key issues they must address if the transition is to be successful. These include how to create a competitive generation sector, how to ensure adequate transmission capacity, how to create a functioning ISO with viable trading arrangements, how to transition to customer choice, how to regulate the remaining transmission and distribution assets, and how to ensure reliability.

Transitioning Generation

As long as generation remains part of a vertical utility and is subject to rate base treatment, there cannot be a fully competitive supply sector because these assets will benefit from cross subsidies not available to merchant generators. Options for opening up the generation sector include ordering the utilities to separate generation functions into a subsidiary company (and regulating activities between the newly created subsidiaries with market affiliate rules), encouraging the utilities to sell off generation assets (known as divestiture), requiring utilities to give up control of generation by auctioning off rights to blocks of generation capacity, or setting specific limits on the market share any one generation entity is allowed to own and control. Forced divestiture (i.e., without a utility's agreement) is legally problematic since it could be considered a "taking" under the U.S. Constitution.

An additional generation transition issue is how to deal with generation assets that may not be well suited to a competitive environment. Examples include nuclear generation (due to its inability to ramp up and down and the need to cover capital costs that may have been allocated to customers over a long number of years), QFs (due to laws that require utilities to take QF output), and certain types of renewable generation (because they are potentially more expensive to build but less harmful to the environment, and may operate intermittently). Often regulators must create a category for nuclear generation and QFs that is called regulatory "must take." This requires the system operator to take the output regardless of price. Mechanisms must also be established for generators that must run due to locational transmission issues that prevent power from being brought into a demand zone. These facilities are called "must run." Regulators must then come up with a mechanism to allocate costs of must take and must run generation to consumers. Regulators or legislators may also wish to encourage development of renewable generation for the societal benefit of a cleaner environment. A common solution is a renewable portfolio requirement. A portfolio requirement sets a percentage of generation that each market participant must obtain from renewable energy. An alternative is to offer tax credits or other economic incentives for building renewable generation.

Creating a Robust Transmission Market

As we have seen earlier in this book, creating supply choice does little to opening a competitive market without also ensuring adequate transmission access so that suppliers can easily compete in all areas of the market without being stymied by transmission congestion. Given current and pervasive opposition to the construction of transmission in this country, this is a significant issue. It appears that the best solution may be the creation of transmission-only companies that are regulated under traditional rate-of-return methods. These companies can significantly expand earnings only by expanding rate base (i.e., by building new transmission), which should strongly incite them to work with local communities to come to acceptable solutions for the construction of new transmission. Although this is a relatively new concept in the U.S., it seems to be having some early success.

A related issue is how to handle existing long-term transmission agreements. Such agreements, if they control a significant amount of transmission capacity, will prevent a competitive market from evolving. If these contracts are held by utilities, regulators can order them to make the capacity available to the marketplace. However, if these are held by municipal or federal entities, as they are in many regions, there is nothing to do but attempt to negotiate.

Finally, in areas where transmission congestion is prevalent, market participants must often pay transmission congestion charges that are in excess of the standard transmission costs. Since this creates a significant cost uncertainty, it is important to create mechanisms that market participants can use to lay off this risk. This is done by creating a mechanism called FTRs, or financial transmission rights. FTRs are often auctioned once a year and allow buyers to lock-in costs associated with a specific transmission path. There is also a need for long-term management of congestion price risk to allow new generation projects to develop without excessive uncertainty. Thus ISOs are beginning to consider creating long-term transmission rights.

Creating an ISO

There is no way around it – there must be an independent unbiased system operator for a competitive market to function. Creating an ISO requires cooperation between the states and FERC, since ISOs are FERC-jurisdictional yet the utilities that will give up the system operations function are state-jurisdictional. And if utilities are not willing to voluntarily turn control of their systems to the ISO, regulators must be willing to order them to do so. As we have discussed elsewhere, the structure of the ISO's trading arrangements are critically important. If the rules are flawed, market consequences can be severe. Fortunately, we now have some good operating experience from various ISOs in the U.S. and around the world, and our knowledge of what works and what doesn't is expanding rapidly.

Transitioning to Customer Choice

Transitioning to customer choice is perhaps even more critical than all the other issues we've discussed. The efforts to create competition are a moot point if customers choose not to participate. Critical issues include customer education, avoiding the temptation to give customers the option to stay with the utility at fixed prices that do not reflect market fluctuations, and letting the market develop sufficiently before dumping customers into the fray. Many observers now believe that it makes most sense to first allow large customers to choose, then transition in customer choice to smaller customers as markets mature.

Continued Regulation of Transmission and Distribution

Regulators must continue to regulate the monopoly functions of transmission and distribution. Options range from traditional cost-of-service to various forms of incentive ratemaking. It is critically important to vigilantly monitor any set of rules for signs that it may create barriers to competition.

Ensuring Reliability

One of the hottest issues in today's environment is how to ensure reliability in a competitive marketplace. Reliability breaks down into four issues – ensuring adequate supply on a long-term and short-term basis, ensuring adequate reserves, ensuring adequate transmission, and ensuring sufficient investment in distribution construction and maintenance. The latter issue can be handled by regulation of the distribution monopoly just as it is under traditional markets. Ensuring adequate supply and reserves in the short term is an issue of trading arrangement design. Ensuring long-term supply and transmission is a more thorny issue. The general concept is that as supply drops, prices will rise. Higher prices encourage new generation or transmission, which enhances supply, causing prices to drop back down. And if you believe in competitive markets, it all works out. Unfortunately, there are often numerous barriers to new generation – NIMBY activists who block construction, uncertain returns given volatile electric prices and lack of capital for speculative construction. And if generation takes five years to build, what happens to customers in the meantime?

A number of options have been discussed and in some cases implemented. Initially many markets attempted to simply let the market work it out. To do so, however, requires mechanisms that allow customers to respond to price. If prices get too high, and the market structure permits customers to respond, they will likely use power in different ways or sign long-term contracts with generators to guarantee prices. However, concerns about rising prices and/or falling reliability have generally led policymakers to conclude that the unique nature of electricity may not be well suited to total dependence on markets. Most regions now give ISOs or state agencies the power to forecast future generating capacity needed to ensure reliability five to ten years into the future. The ISOs or state agencies then allocate a capacity responsibility to each load serving entity based on that entity's market share. The load serving entity is then obligated to make sufficient long-term supply arrangements to cover its allocated responsibility or face penalties. This then provides a long-term guaranteed revenue stream for generators, resulting in construction of sufficient capacity.

Settlements

As you might imagine, the shift to numerous market participants and multiple energy, reserves and capacity markets creates the need for additional accounting and tracking of transactions to ensure money flows to those providing necessary services. This important function is provided by a centralized entity that tracks power transactions and determines who is owed (and by whom) for the thousands of daily transactions. For wholesale transactions this is typically handled by the ISO.

What you will learn:

- The recent history of electric market restructuring in the U.S.
- The current status of restructuring in the U.S.
- What different states have done and are doing
- What has been done elsewhere in the world
- Key issues that have arisen during the restructuring process

SECTION TWELVE: THE HISTORY OF ELECTRIC MARKET RESTRUCTURING

The door to electric market restructuring cracked open in the United States in 1978, when a change in federal law allowed private owners of cogeneration units to sell power into utility grids. Prior to this time, the U.S. had no non-utility owned generation (other than generation owned by groups of utilities in power agencies or generation owned by federal power agencies). But it wasn't until the 1990s that we began to see the actual implications of restructured markets. This is when issues such as merchant generation, transmission access, ISOs, and customer choice programs came to the forefront. Due to a shared regulatory jurisdiction, market restructuring requires cooperation from both federal and state regulators. This has worked well in some cases and not so well in others. Because of the states' role in regulating utilities, the status of market restructuring varies widely across the U.S. Some states have gone well down the path to competition, while others have held on to traditional regulation. In this section we will look first at the history of federal government deregulation, then move on to a discussion of the states.

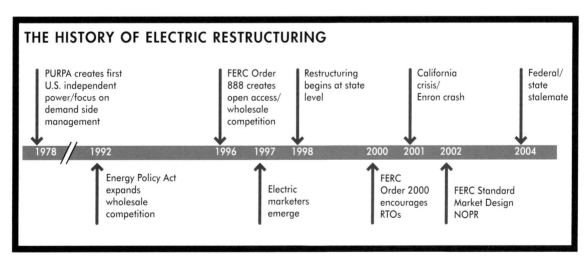

SECTION TWELVE: THE HISTORY OF ELECTRIC MARKET RESTRUCTURING

Federal Restructuring

The First Step Towards Independent Generation – PURPA of 1978

The first move towards restructuring of U.S. electricity markets occurred in 1978, when Congress passed the Public Utilities Regulatory Policy Act (PURPA). Passed during the 1970s energy crisis, PURPA required, among other things, that electric utilities connect with and buy capacity and energy from any facility meeting the criteria for a qualifying facility (QF). PURPA also required that the utility pay for the purchased power at the utility's own "avoided cost" of production. The avoided cost refers to the price the utility would have otherwise paid to generate that power itself. The criteria for becoming a QF included the requirement that the facility be either a cogeneration facility with a certain defined efficiency level or powered by renewable fuels. Other criteria included limits on utility ownership percentages and size of the units (see What is a QF? on page 48).

Although Congress mandated that utilities purchase QF power, the details were left to the states to implement. Utilities were left to set interconnection requirements (rules about how QF facilities could be connected to the utility system), determine how to calculate avoided costs, lay out the contractual terms for operation of the facilities, determine the mechanism and cost responsibility for any necessary system upgrades to accept the power, and determine the appropriate gas rate[1] to charge these facilities. Some states were favorable to QF power, creating beneficial payment schedules and pushing the utilities to foster QF connections. Others were less favorable and allowed utility rules that impeded development of QFs. Thus, in some states like California and New York significant QF development occurred, while in others very little developed.

Fostering Wholesale Generation Competition – the Energy Policy Act of 1992

Following PURPA, the concept of independent power producers (IPPs) became a trend in the utility industry. Due to issues with construction of nuclear units and other big utility projects, regulators in some states began to favor units built by IPPs which were contracted to the utilities for the life of the unit. By 1992, IPPs were building 60% of the new capacity in the U.S. But opportunities were limited since sales could be made only to the incumbent utility. Congress moved to address this issue with the Energy Policy Act of 1992. Provisions of this legislation include:

---

[1] QFs, because they are essentially competing with generators, often receive natural gas service rates that are comparable to those paid by generation customers.

- A new category of generator was created called exempt wholesale generation (EWG). These generators were not restricted by size or fuel type as in PURPA, were not subject to state jurisdiction, but also had no guaranteed right of sale at the avoided cost.

- Owners of EWGs were exempt from the provisions of PUHCA (the law dating from the 1930s that restricted ownership of utility assets by holding companies).

- Owners of EWGs had no rights to sell directly to end-use customers.

- FERC was authorized to compel utilities to transmit a third-party's power across the utilities' lines (wheeling).

- Utilities were ordered to use integrated resource plans (IRP) in determining future resource requirements, to file such plans with their state regulator, and include in the filings plans for implementation of the IRP.

- FERC was prohibited from ordering or approving retail access or choice programs.

- State regulators were required to implement rate structures that would remove profit disincentives for utilities to support demand side management programs.

The key points to the Energy Policy Act which laid the groundwork for restructuring are that it created the legal framework for IPPs to sell to someone other than the local utility, it gave FERC the authority to order wheeling, and it put in place the IRP process that required utilities to consider purchased power as well as construction of power plants.

While the Energy Policy Act furthered the movement towards competitive generation, it was not enough. Provisions in the bill limited wheeling by requiring a case-by-case analysis of wheeling requests to ensure, among other things, that reliability was not unreasonably impaired, that the interests of the transmitting utilities' own customers were not unduly disadvantaged, and that existing contracts were honored. And, by preventing generators from selling directly to end users, the bill ensured utilities remained the sole buying customers.

Following the Energy Policy Act, industrial customers began pushing for direct access to purchase electricity, while generators and utilities became increasingly antagonistic as to whether utilities were providing fair access to wheeling; and when they did, whether they were charging fair rates for wheeling services. IPPs accused the utilities of making it difficult to wheel. Utilities became increasingly concerned with stranded costs (see box on page 147) and the possibility that power from outside their area might be brought in more cheaply than power could be provided from some of their own more expensive units.

Furthering Open Access Transmission – FERC Order 888

Seeing the problems the existing policies were creating, and mindful of the success of open access policies in the natural gas industry, FERC opened a proceeding to look into the issues of transmission access, transmission pricing and stranded costs. This resulted in FERC Order 888, issued in April 1996. Order 888:

- Required that all transmission owners subject to FERC jurisdiction provide wholesale transmission services to all parties under the same terms and conditions that they provide service to their own generation, with the exception that utilities were able to reserve transmission for service to their own native loads (meaning end users of the utility).

- Required utilities to functionally separate generation, transmission, power control, and distribution activities.

- Identified six ancillary services that utilities must provide in adjunct with transmission service and allowed utilities to develop rates for these services.

- Found that if stranded costs are caused by departing wholesale customers, the utility could recoup these costs from the departing customers, providing the utility first tried to mitigate these

ORDER 888 PRINCIPLES FOR APPROVAL OF ISOs

The ISO:

1. Must have a fair and non-discriminatory governance structure.
2. May not have financial interests in any market participants.
3. Must have a single, open-access tariff for the entire area served by the ISO.
4. Is responsible for system security.
5. May control system dispatch for pool or bilateral arrangements.
6. Can manage transmission constraints.
7. Has incentives for efficiency.
8. Has pricing mechanisms for transmission and ancillary services that promote market efficiency.
9. Must post transmission availability in real time on electronic bulletin boards.
10. Must coordinate with adjacent control areas.
11. Must have a dispute resolution process.

costs (e.g., finding new customers to replace them).

- Encouraged, but did not require, utilities to create Independent System Operators (ISOs) and laid out criteria for FERC approval of them.

Order 888 put in place a number of key conventions that still drive the way the electricity industry operates today. The Order required open access tariffs (often called OAT tariffs) whereby utilities must treat other parties' transactions in the same manner they treat power transactions performed internally. It required the utilities to functionally create separation between their generation, transmission, power control and distribution departments (although FERC did not have the authority to order the actual break-up of the vertical utility). It introduced the concept of ancillary services as separate and distinct and allowed utilities to charge for providing them. And, much to the relief of the utilities, it embraced the concept that stranded costs should be recoverable.

> **STRANDED COSTS**
>
> One of the most contentious issues regulators must consider during restructuring is how to handle stranded costs. Stranded costs are utility costs associated with assets acquired under prior regulatory rules that are in excess of the market value of these assets once the market is restructured. An example is a high-priced power plant that will no longer be run once lower-cost merchant power comes into a restructured marketplace. Utility shareholders will argue that the above-market costs should be paid by consumers since the unit was initially built under the assumption that the utility was responsible for serving all loads in its territory. Companies attempting to compete with utilities will argue that providing full cost coverage for stranded assets makes it impossible for competitors to compete effectively and thus deprives consumers of the benefits of competition. A key job for regulators is to accurately identify stranded costs, and then collect them in a manner that does not impede the development of a competitive market.

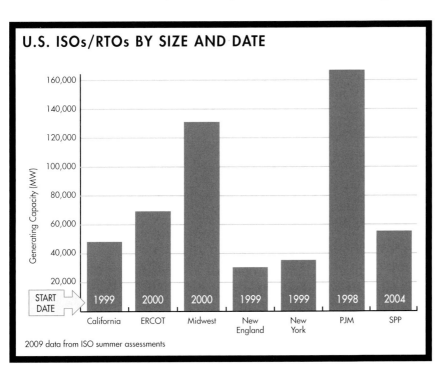

U.S. ISOs/RTOs BY SIZE AND DATE

2009 data from ISO summer assessments

147

Although this applied only to wholesale issues, it gave utilities a good precedent to quote when the discussion came up later in state restructuring proceedings. Order 888 also encouraged, though it did not require, the formation of ISOs and laid out 11 principles required for approval of ISO tariffs (see box on page 146).

Wholesale trading grew quickly after Order 888 was issued – from approximately 100 million kWh in 1996 to close to 4,500 million kWh in 2000[2]. However, until states began restructuring, the only buyers were still the utilities. Restructuring in some key states including California, New York and Pennsylvania began in the late 1990s, opening the door for marketers and generators to sell directly to end-use customers.

Encouraging Formation of Regional Transmission Operators – FERC Order 2000

Order 888 forced utilities to functionally separate generation, transmission, distribution, and power control, and encouraged utilities to turn the power control function over to ISOs. A number of regions moved forward voluntarily with ISO formation including California, New England, New York, and PJM (Pennsylvania, New Jersey, and Maryland, plus areas of other neighboring states). The northeast ISOs were all evolutions of the utilities' power pools, and thus started with an existing structure in place. California combined the power control functions of Southern California Edison and Pacific Gas and Electric to create a new statewide marketplace. Even with this movement in some regions, it became clear that fragmentation in transmission rules was causing a significant barrier to electricity trading. Marketers wanting to move power were often faced with rate pancaking (paying multiple rates to multiple transmission entities) and/or were faced with different sets of rules across different transmission systems – either of which could make transactions unviable. To create a larger marketplace (like the one that has worked so well in the natural gas industry), FERC needed to find a way to eliminate the wide disparities in rules and rates in the various utility and ISO tariffs that resulted from Order 888. Even a statewide ISO like California or New York could be problematic since historical power flows have occurred between these and neighboring states operating under different rules.

FERC concluded that a competitive market could be fostered by moving to system operators that created uniform markets across larger regions, and coined the term Regional Transmission Operator (RTO). Like an ISO, an RTO would handle all the power control, transmission access and ancillary services market functions but would

---

[2] It should be noted, however, that later events revealed that some of these transactions were sham transactions used by marketers to boost apparent revenues.

UNDERSTANDING TODAY'S ELECTRICITY BUSINESS

do so over a wide geographical region which reflects physical market realities of the regional grid. FERC's vision is that the U.S./Canada marketplace should evolve to a system of a limited number of large RTOs.

In an attempt to push the industry towards its desired model, FERC issued Order 2000 (actually issued in late 1999) which defined the minimum characteristics and functions for an RTO, set forth a voluntary collaborative process intended to lead to the formation of RTOs, and required jurisdictional utilities to make filings with proposals for creating and joining RTOs that would be functional by December 2001 (or to tell FERC why it wasn't possible to meet that date and what they were trying to do to get there).

Order 2000 was an attempt by FERC to move the market where FERC wanted it to go. Yet it was unclear whether it even had the authority to order state-regulated utilities to consolidate their operations into RTOs. FERC's position was clear: "As a result of this voluntary approach, we expect jurisdictional utilities to form RTOs. If the industry fails to form RTOs under this approach, the Commission will reconsider what further regulatory steps are in the public interest.[3]" A number of filings ensued, with over a dozen proposals for RTO formation. As a result, a number of the existing ISOs were certified as RTOs and a new RTO, Southwest Power Pool (SPP), was approved in 2004.

Several regions raised opposition to the movement to RTOs, fearing loss of state control and possible loss of control over lower-priced generation resources. This attitude was especially prevalent in the Southeast, Northwest and parts of the Rocky Mountain West. Two other factors played against the movement towards FERC's vision – the California energy crisis of 2000/2001 and the Enron crash in late 2001. Suddenly many states and other entities that had been willing to consider restructuring got cold feet. And those that opposed FERC's vision had ready ammunition to argue against it.

The Current Stalemate

Realizing its vision wasn't moving forward, FERC made an additional attempt to push the marketplace forward by initiating a new rulemaking proceeding in July, 2002. The proceeding called Standard Market Design Notice of Proposed Rulemaking (known as the SMD NOPR, or simply SMD) proposed more forceful measures to implement standard market designs across the U.S. Under SMD, FERC proposed standard designs

---

[3] FERC Order 2000, p. 4.

for RTO markets and uniform transmission tariffs across each region. FERC apparently underestimated either the strength of opposition to SMD or its ability to push things forward in the face of opposition. After legislators from Southern and Northwestern states threatened to hold up FERC's budget authorizations and then proposed bills that would explicitly limit FERC's authority to impose SMD, FERC backed down.

In 2003, FERC issued a new white paper that indicated a reduced emphasis on uniformity with greater tolerance for regional variations. It also suggested that transition periods could be as long as ten years. FERC's attempts to create a uniform market structure across the U.S., or even to foster a rapid transition to regional uniform markets appear at least temporarily stymied. This leaves us with a fragmented market characterized by seven functioning ISOs/RTOs, three additional proposed RTOs with uncertain dates for implementation, and other regions with no ISO/RTO plans.

Some observers believe that over time the market will evolve itself towards FERC's vision, while FERC continues to do what it can to prod participants in this direction. Both California and Texas have plans to move their market structures in the direction proposed in SMD. And MISO and PJM are moving forward with plans to create a uniform market structure across a wide area of the Eastern and Midwest markets. Other discussions have included consolidation of TVA, New York ISO, ISO New England, PJM ISO and Southwest Power Pool (SPP) market structures, although this proposal is not active. If market participants do ultimately move in this direction, it would be a significant step as this single market would cover as much as 45% of U.S. demands. Although the amount of the market covered by ISOs continues to grow, market participants have no choice but to accept the necessity of operating in an environment of fragmented markets.

State Restructuring

While it is FERC's role to get wholesale trading, transmission and system operation markets working, it is the role of the states to consider retail competition and breakup of the vertical utilities. Some states have chosen to do nothing, others have moved aggressively into restructuring the role and business structure of the utilities. In many cases it is the high price states that have taken the lead, given pressure from industrial customers who desire access to lower-cost sources of electricity. The first state to implement retail access was Rhode Island in 1997. By 2006, 20 states plus the District of Columbia have implemented some form of retail competition.

Key issues for the states include:

- Which market structure to adopt.
- Which trading arrangements to adopt.
- Whether to support movement to an ISO or RTO.
- How to foster the separation of the vertical utility functions.
- Whether to allow retail access, and if so, how.
- How to continue regulation of the continuing monopoly utility function.

In general, states that have chosen to restructure have done it in one of three ways — transitioning existing power pools into ISO structures (New England states, New York, PJM states), creating new state-wide or regional ISOs (California, Texas and portions of the Midwest), or implementing restructuring without putting in place new competitive wholesale market structures (the rest).

We discussed market structures and trading arrangements in Section Nine, as well as ISOs and RTOs earlier in this section. Following is a discussion of the remaining key

issues faced by states and how they've been handled to date. Suffice it to say that no two states have implemented restructuring exactly alike, and each state is a unique case.

Separating the Vertical Utility Functions

As we have discussed, it is necessary to separate the utility functions of generation, transmission, system operations, and distribution to provide opportunities for competition. Absent some breakup of utility functions, non-utility competitors will always be looking in from the outside and will not be able to compete on a level playing field. Actions states have taken range from simply requiring utilities to create new separate departments (with affiliate rules to define behavior between departments) to strongly encouraging utilities to divest of generation and in some cases transmission assets. Another option is to have utilities move their generation assets to separate subsidiary companies.

Allowing Retail Access

In general, the decision for retail access comes down to allowing all customers to choose, allowing none to choose, or defining specific classes of customers that are eligible to choose. A popular solution is to allow only large commercial and industrial customers to choose. Another is to allow all customers to choose competitive supply, but also to continue a regulated default supply service from the utility, often with capped rates. Some states have skirted the issue by allowing customers to leave utility supply on a case-by-case basis that requires specific commission approval or by experimenting with virtual access. Virtual access offers customers the option of having the utility buy power at market rates for the customer rather than receiving a traditional average utility supply cost.

Continued Regulation of the Monopoly Function

The mechanism for continued regulation of the monopoly function is often one of the most contentious issues and can determine everything about the attractiveness and success of retail access options. The issues to be considered are discussed in Section Eleven, but it is worth reiterating that the devil is in the details and seemingly minor decisions such as how meter data will be transmitted among market participants can have huge impacts. Many states that have implemented retail choice continue to work on such details, and it is important to remember that restructuring is an ongoing effort, not a one-time event.

The California Experience

It is unlikely that any reader of this book on the electric industry has not been exposed to some media coverage of the California energy crisis of 2000-2001. And most have probably heard as much as they care to on the subject. If you are well versed in California's experience in electric restructuring, you are encouraged to skip the following discussion. If, on the other hand, you do not fully understand what occurred in California, read on for a great lesson in politics and market dynamics.

California implemented electric deregulation on April 1, 1998. In the first years (1998-1999) wholesale electric prices dropped. So much so that some market participants in the Pacific Northwest accused California of creating a market structure that would purposely drive prices down to the benefit of Californians and the detriment of surrounding states that export large volumes to California. Retail access markets were active and things seemed to be going well.

Suddenly everything changed. In the summer of 2000 high market prices began occurring in peak hours, and by fall prices were high in almost all hours. Wholesale power costs rose from $7 billion in 1999 to $27 billion in 2000. The utilities[4] were unable to pass these costs through to their customers due to the structure of the market. Thus the utilities were put in the position of buying power at an average wholesale price of $317/MWh in December 2000 while reselling it to their customers at rates in the $45/MWh range. The utilities had nowhere to turn to make up the revenue shortfall, and soon could no longer pay the generators for power. Power was already short due to a low hydro year and numerous generators off-line for maintenance. (It now appears that some generators may have purposely removed their facilities from the market because they were not getting paid, or in some cases to drive prices higher.) By January 2001, the California ISO could no longer procure enough generation to serve loads and was left with no alternative but to implement planned rolling blackouts to bring supply and demand in balance.

The State of California was forced to step in as the only creditworthy buyer able to procure supply on behalf of the utilities. The state signed long-term agreements with numerous generators to assure supply at prices and terms that were highly favorable to the generators. Given market conditions at the time, however, the state had little choice but to sign them. These contracts quickly proved to be an embarrassment as

---

[4]By utilities we mean Pacific Gas and Electric and Southern California Edison. The situation was different for San Diego Gas and Electric and the municipal utilities.

spot prices – driven by increased supply and falling demand – fell well below the fixed contract rates and the state was forced to sell excess power at a loss (although in fairness to the state, it should be noted that the existence of the agreements contributed to stabilization of the market and perhaps helped to drive market prices down). The state found itself with billions of dollars in general fund expenditures for electricity that was not recovered in electric rates. Pacific Gas and Electric was forced to declare bankruptcy and Southern California Edison nearly did as well.

California suspended retail access to prevent customers from leaving the utilities' supply (to avoid paying for the high-priced state contracts) and went back to the regulated utility model. By 2003, the utilities were back in charge of generating and/or purchasing supply for all customers (except for grandfathered direct access customers) under a cost-of-service regulation model, and by 2004 the California Public Utilities Commission (CPUC) allowed San Diego Gas and Electric and Southern California Edison to purchase new generating assets within the regulated utility. Meanwhile, California customers who were already paying some of the highest rates in the country were hit with new rate increases to cover the cost of paying off past debts and of covering the high-priced state contracts. The state accused a number of market participants of purposely manipulating the market to raise prices, and information disclosed in subsequent proceedings seemed to indicate that some had indeed done so. FERC ordered a number of refunds and the state came to settlements with other parties.

Meanwhile, Governor Gray Davis was swept out of office in a recall election caused in part by public perception of his failed handling of the crisis and was replaced with former movie star Arnold Schwarzenegger. By early 2006, the three investor-owned utilities were back in the business of providing monopoly service to all but a dwindling group of direct access customers who were grandfathered from earlier years.

Many who oppose electric restructuring argue that California shows why it is a flawed concept. Others who support restructuring simply believe that it was the California market structure that was flawed – along with a good bit of bad luck. And many who don't support one position or the other are simply scared away from restructuring by the outcome. So why did California's experiment turn out so badly?

California's experience was the result of a series of unfortunate circumstances, many of them self-inflicted. AB 1890, the California bill that set the market structure was passed unanimously in the state legislature. This perhaps led to much of the problem since the bill included numerous provisions attempting to appeal to everyone. Key factors included:

- A supply dependence on hydro power, which varies significantly from year-to-year based on weather patterns.

- A low hydro year in 2000 coupled with rising natural gas prices.

- A lack of new generation or transmission built in California for ten years prior to restructuring.

- An incredibly complex market structure that included bilateral trading for non-utility entities, a requirement that utilities trade through a centralized auction-based exchange that was separate from the ISO, and an ISO that used auction-based markets every hour to acquire ancillary services and to allocate transmission during times of transmission congestion. With day-ahead, hour-ahead, and real-time markets this resulted in over a dozen auctions for energy and capacity for every hour of every day.

- A market design that resulted in the utilities buying all their supply at spot prices. It is estimated that upwards of 80% of California power was being bought at spot market prices whereas in other markets 80% was bought at long-term prices. And the spot prices were determined in auctions where the most expensive unit used during the hour set the market price that was paid to every supplier. So 80% of the state's power was bought at the marginal price.

- A retail market structure that provided end users a fixed rate no matter what happened in the wholesale marketplace. Thus, there was no demand response even as wholesale prices climbed to ridiculous levels. Many analysts believe that the California crisis would never have occurred had end users been exposed to wholesale prices – many users would have simply reduced energy use at those price levels and the problem would have corrected itself.

- Corporate culture among market participants that seemed to accept the concept that any action that resulted in profits was good, along with a lack of market protections by regulators. Both FERC and the CPUC seemed to enter deregulation believing that the market would take care of any attempts at manipulation. But a complex market structure is rife with opportunities for what traders call "arbitrage," and the California ISO was not given the teeth to police activities which were clearly designed to profit traders in questionable ways.

The California crisis has been studied extensively across the country, and one can only hope regulators and legislators have learned a number of lessons about designing restructured markets. Meanwhile, California may soon try again by requiring direct access options for industrial customers.

SECTION TWELVE: THE HISTORY OF ELECTRIC MARKET RESTRUCTURING

Restructuring in Other Countries

In these days of increasing globalization, it would be foolish not to look beyond our borders at restructuring efforts in other parts of the world. Concepts and technologies developed elsewhere have the potential to rapidly change the U.S. marketplace. While we won't go into detail on the world electric marketplace, it is important to be aware that electric restructuring is moving rapidly in many places throughout the world, and that the U.S. is no longer on the forefront of restructuring efforts. The European Union plans to bring retail choice to all customers over the next few years, and full retail customer choice with the equivalent of ISOs has been in effect in Australia, New Zealand and Britain for a number of years. Other countries such as Chile and the Nordic countries of Norway and Sweden have operated competitive generation markets for over a decade. And the Canadian provinces of Alberta and Ontario have implemented competitive markets. Thus electric restructuring will continue to be a force worldwide. While the movement towards restructuring has slowed in the U.S. in recent years, it is sure to continue, and those states that have moved well into restructuring will probably never go back to a world of vertical utilities.

What you will learn:

- How electric supply and demand fluctuate
- How wholesale electricity prices are set
- Why wholesale prices are so volatile
- How the wholesale marketplace works
- How the retail marketplace works

13

SECTION THIRTEEN: MARKET DYNAMICS

Ultimately, all activities in the marketplace are dictated by the end user, who will purchase electricity only so long as it is financially feasible and emotionally rewarding to do so. One of the major sources of inefficiency in the electricity marketplace is the fact that consumers have generally been shielded from the actual costs of their end-use decisions because most pay an average cost. As described by Sally Hunt, an expert on electricity competition[1], if we ran clothing stores the way we run the electricity market we would offer Armani suits and Gap jeans at the same average price. Of course, there would be a significant migration to Armani suits since they would be available at the same price as an outfit from the Gap. Operating the way electricity traditionally has worked, manufacturers would simply respond by manufacturing more Armani suits and the average price for all consumers would rise. This works great as long as you have a regulator who can force everyone to pay for the more expensive product (i.e., Armani suits) in an average price. But once you let anyone have direct access to buy directly from the Gap, at Gap prices, the whole system falls apart. We tell you this story because while historically our regulatory mechanism has served up average prices, significant portions of the U.S. marketplace can now negotiate pricing with the electric equivalent of the Gap. This reality will continue to fundamentally change markets regardless of the outcome of FERC's Standard Market Design or other regulatory initiatives.

In many parts of the U.S. we now have business structures in which wholesale power is widely traded and where significant amounts of power are traded at market prices (as opposed to regulated prices). In others, the hold of the vertically-integrated monopoly utility is still so strong that no competitive marketplace exists. Thus there are wide discrepancies across the U.S. in the maturity of wholesale markets. The same dichotomy exists to an even greater degree in retail markets. In a few states, most customers purchase power priced at market rates. In others, all customers still buy supply from a monopoly utility under traditional utility cost-of-service pricing. And many states have a mix where some customers are exposed to market prices and other aren't. As you can see, market dynamics vary widely from region to region.

---

[1] Sally Hunt, *Making Competition Work in Electricity*, page 78.

| REGIONAL MARKETPLACES |||
Region	Wholesale Competition	Retail Competition
Northeast/Mid-Atlantic	High	High
Southeast	Low	Low
Midwest	Medium, growing	Medium, slowly growing in some states
Rocky Mountain	Medium	Low
Northwest	High	Low
California	High	Low
Texas	High	High

Supply and Demand

Because regulation has traditionally insulated customers from market pricing, electricity markets have not historically followed standard economic principles. The general paradigm has been supply will be built to meet forecasted demand, regardless of cost. And demand is based on demographics, business activities and weather patterns – again, regardless of cost issues. The other key factor about electricity demand is that time-of-use is critically important since peak demand drives much of the cost of the system, and capacity must be planned to meet the overall market peak even though that may occur only a few hours out of the year. Some utilities have moderated demand growth through demand side management (DSM) programs that encourage customers to enhance energy efficiency or move demands to off-peak periods. But in general, we have traditionally built power plants and transmission lines in step with forecasted demand increases.

In the long term, supply has historically been driven by demand forecasts. Once a utility's forecast indicates that a region's available supply is getting close to demand, the utility will request new resources in its integrated resource plan, regulators will approve either the construction or purchase of new resources, and voila, supply is increased. But in the short term, increasing supply is more difficult since it generally takes at least two years (and often much longer) to plan, design, permit, and build a new power plant. Thus there are really two sets of supply/demand issues to consider: long-term issues which relate to decisions to build new infrastructure, and short-term issues which determine whether there is enough supply to cover customer needs for today and tomorrow.

The way that the balance of supply and demand in a specific market region is evaluated is by looking at the reserve margin. The reserve margin is calculated as the total supply capacity in a region (supply can include generation plus available firm imports) minus the peak demand, divided by the peak demand. For instance, if a marketplace had 12,000 MW of supply and 10,000 MW of peak demand, the reserve margin would be: (12,000 − 10,000) ÷ 10,000 = 20%. Essentially, we are calculating how much extra supply is available to a given region. This is important for extreme weather situations or in the event any of the total supply should suddenly become unavailable. Markets with reserve margins of less than 15% are considered tight. Those with margins between 15% and 20% are considered balanced, and those with margins greater than 20% are often considered oversupplied.

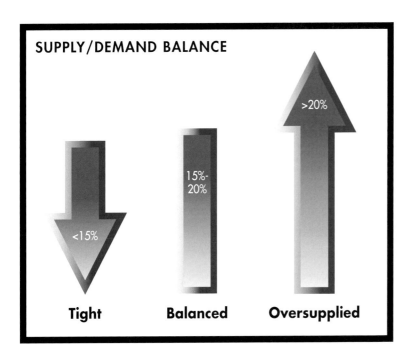

Short-term Supply and Demand

The short-term supply/demand balance is driven simply by projected demand and the supply available to meet it. The marketplace will generally look at this on a seasonal and monthly basis (since blocks of power are often traded and/or priced seasonally and monthly) and then again very closely in the day-ahead (since this is when the system operator will schedule actual units and transmission lines). In the short term, demands are driven largely by weather (hot weather in summer driving cooling loads and cold weather in the winter driving heating loads) as well as business activity (often determined by what day of the week it is). Short-term supply is driven by the availability of generation units, transmission and firm power imports.

A number of factors can impact unit availability including maintenance needs, environmental restrictions, fuel availability, and weather patterns (for renewable resources such as wind and solar). In a competitive marketplace, additional factors such as contractual conditions, tariff provisions and the behavior of market participants also come into play. On an hourly basis, unit availability is also impacted by start-up and ramp-up times. Units that haven't been started may not be available for a given hour if the time it takes to get them on-line safely is longer than the hour in which they are needed.

Transmission line availability is also an important factor when determining available supply. This is affected by weather (hot weather reduces capabilities), maintenance needs and use of the lines by other market participants. If supply/demand is tight for a given day or hour both market prices and system reliability may be impacted.

Long-term Supply and Demand

In the longer term (greater than one year), demand is driven largely by demographics and business cycles. Long-term supply is affected by construction of new units and/or transmission lines, retirement of units and in some markets availability of hydro power (driven by weather patterns). Construction of new units can be impacted by capital availability, regulatory decisions, environmental restrictions, willingness of buyers to sign long-term power purchase agreements, and market participants' perception of future profit opportunities (in competitive markets).

The Current Supply/Demand Situation in the U.S.

As of late 2009, the supply/demand situation appeared to be stable. In years priors to 2008, summer peak demands had grown dramatically and many regions felt the need to plan for significant generation construction. But the economic downturn that began in mid-2008 coupled with increasing adoption of demand side management dramatically changed the outlook. It now appears that current generation combined with construction in progress, planned renewable generation development, and expanded economic demand response programs may be sufficient to reliably cover demands for the next decade. But as with most forecasts, uncertainty reigns. Key factors that will influence demand growth include how quickly the econ-

---

SHORT-TERM SUPPLY/DEMAND FACTORS

Supply

- Units with long start-up/ramp up times
- Units out for maintenance
- Environmental permit restrictions
- Fuel availability
- Weather impacts on renewables
- Contractual and tariff provisions
- Actions of generation owners
- Transmission line availability
- Availability of firm import power

Demand

- Weather
- Business activity
- Availability of demand response

LONG-TERM SUPPLY/DEMAND FACTORS

Supply

- Ease of new generation/transmission construction
- Retirement of units
- Long-term weather patterns

Demand

- Regional demographics
- Business cycles

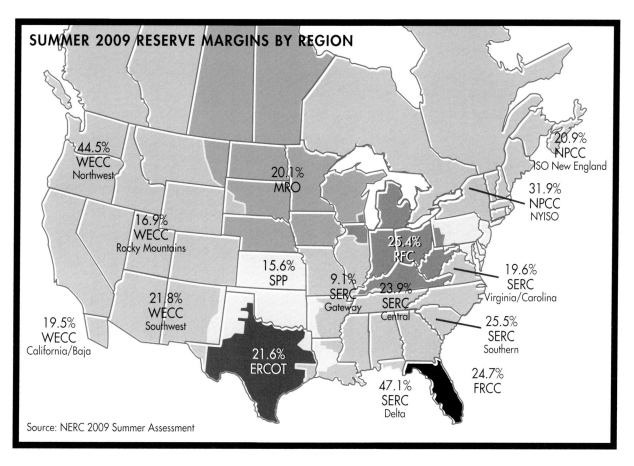

Source: NERC 2009 Summer Assessment

omy recovers and whether current efforts to enhance energy efficiency will result in long-term demand reductions.

Pricing

Various factors influence the price of electricity at any given location at any point in time. Generally speaking, market-based prices are determined by market perceptions of the supply/demand balance. But in the electricity business things are more complex because some prices are set by regulation, not by the market. In this discussion we will consider only market-based pricing. The principles of regulated prices, which in the wholesale market are generally cost-of-service pricing, are covered in Section Ten. Market-based electricity prices are usually determined in one of two ways – through bilateral transactions or through centralized bidding. Bilateral transactions usually occur on the phone with two individuals negotiating and agreeing upon a price. For shorter transactions, some traders are moving beyond the phone and are beginning to use electronic exchanges such as the Intercontinental Exchange (ICE). Longer-term transactions are typically negotiated face-to-face. Until the advent of ISOs, bilateral trading encompassed virtually all market-based wholesale transactions and this is still the predominant method for all pricing negotiated further ahead than the next day. Centralized bidding is run by ISOs for day-ahead and real-time markets. These run on either the auction methodology under decentralized trading arrangements or the optimization methodology under integrated trading arrangements (for more details see Section Nine).

Prices in bilateral transactions are driven by each organization's perception of supply and demand and what each party considers a fair price. Centralized bidding is generally driven by the bid price of the last unit turned on in a specific hour to provide the required energy. But keep in mind that a fair bid may be greater than just the variable cost of operation. At some point in the year, owners of generation need to recover not only variable costs but also fixed costs, debt costs and a margin of profit. So during hours where supply is tight relative to demand, bids will be set not just to cover costs in those specific hours, but also to recover the additional items mentioned above. Some peaking units may run only 50 to 100 hours for the entire year. This may not be significant if the unit owner is a utility that has put all the fixed costs and profit of its unit into a revenue requirement to be spread out over the year. It is very significant if that peaking unit is owned by a merchant generation company that has no guaranteed cost recovery.

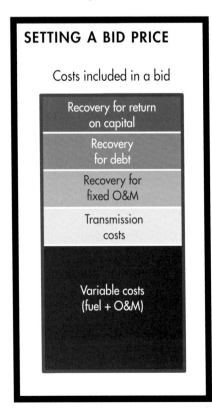

SETTING A BID PRICE

Costs included in a bid

- Recovery for return on capital
- Recovery for debt
- Recovery for fixed O&M
- Transmission costs
- Variable costs (fuel + O&M)

Indexes and Trading Hubs

A basic requirement for a competitive commodity market is open price discovery. This means that all participants have access to information about the market price of electricity at specific locations. Indexes, compiled by buyers and sellers reporting trades and prices to an impartial third party, provide this information. Locations used for indexes are trading hubs which are places where buyers and sellers commonly transfer ownership of electricity. Unfortunately, indexes can be misleading because they depend on accurate reporting from buyers and sellers and also require a large number of transactions to be statistically valid. In a volatile electricity industry, neither can always be counted on. Given revelations in the early 2000s about false price reporting, FERC and market participants have

UNDERSTANDING TODAY'S ELECTRICITY BUSINESS

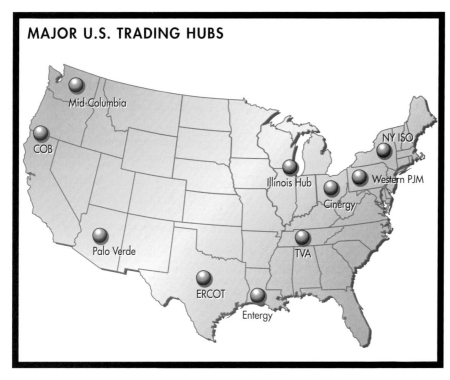

MAJOR U.S. TRADING HUBS

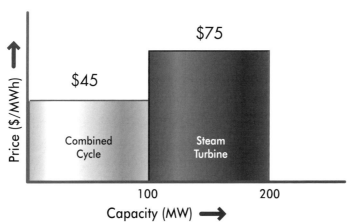

MARKET PRICE VOLATILITY

In the graph above, the marginal cost of generation in the marketplace is $45/MWh as long as demands do not increase above 100 MW. Once demands go above 100 MW, the marginal cost becomes $75. And in an ISO market where everyone gets paid the marginal price, the ISO price for all power becomes $75, even if demand has only gone up from 100 MW to 101 MW. This is why price volatility is to be expected.

worked to improve ways to determine accurate indexes. FERC now monitors markets and can enforce penalties for price manipulation. This is important because many market participants use indexed pricing in power purchase agreements.

Price Volatility

Price volatility, or the movement of price over time, is an inevitable fact of market-based electricity pricing. Since demands rise and fall significantly over the course of the day, and power plants go on- and off-line for maintenance and other reasons, the hourly supply/demand equation is in constant flux. Prices can jump very quickly based on the cost of operation of the next power plant that must be turned on to satisfy an increase in demand (the marginal cost – see box at left).

SECTION THIRTEEN: MARKET DYNAMICS

Many observers say that price volatility in the electricity marketplace is the highest of any commodity that has ever been traded. It is very common on a normal day for electricity prices to go from $25/MWh to $60/MWh. And on a hot summer day where there are transmission line restrictions or units down for maintenance, a price of $150 or even $500/MWh would not be unexpected. In fact, we have even seen prices spike as high as a reported $7,500/MWh in one marketplace! Price volatility is made more extreme by the fact that most electric markets do not have much in the way of a demand response to price. Since customers are being charged average rates that only change once a year, they are oblivious to high wholesale prices – even at a price of $500/MWh (five times higher than a common residential rate). This extreme volatility has resulted in the need for sophisticated risk management techniques which we will discuss in the following section.

The Wholesale Market

A wholesale electric market refers to transactions that occur between two parties, neither of which is the ultimate consumer of the electricity. In some parts of the U.S. a fairly robust wholesale market has evolved over the last ten years. Characteristics of a highly competitive market include many buyers and sellers (a liquid market),

> **MARKET TURMOIL**
>
> In the period 2000-2002 electricity markets were hit with a number of perturbations that threw them into turmoil. Events that occurred include:
>
> - The California energy crisis, which raised wholesale prices to unprecedented levels throughout the western grid.
>
> - The bankruptcy of Enron, the largest trader of electricity in North America, followed by bankruptcies of other large wholesale traders and merchant generators including Mirant, NRG and PG&E National Energy Group. Others avoided bankruptcy but suffered severe financial stress.
>
> - Revelations that numerous companies had engaged in questionable accounting and other business practices.
>
> - Exposure of widespread abuse of tariff conditions, power plant availability rules, and index reporting practices that allowed some wholesale market participants to drive prices to high levels and apparently earn huge profits at the expense of consumers.
>
> These events led to a general questioning and mistrust of those engaged in electricity trading and marketing and, combined with a drop in power prices that caught many merchant generators with inadequate cash flows, led to a huge crash in market values of many companies. A number of entities that had been the largest market players in the wholesale business were forced to attempt to sell assets to stave off bankruptcy and were no longer viable market participants. Over the next few years a number of assets changed hands, new participants emerged and trading continued. After all, trading must go on. Owners of power plants must find customers and load serving entities must find sources of supply. So anywhere the vertical utility can't serve 100% of its loads with its own generation, there will always be some form of electricity trading.

prices determined by market conditions, price transparency, no individual or group of companies with market power, and no barriers to transfer of goods (i.e., no issues with transmission access or capacity, and no regulatory restrictions on transactions). These conditions exist to a degree in various regional markets in the U.S. However, market perturbations between 2000 and 2002 (see box on page 166) have caused concerns, and fragmented regulations and market structures continue to impede the movement towards more liquidity. But restructuring of markets is well underway, and competitive wholesale marketplaces continue to grow in most regions of the country.

Services in the wholesale marketplace include energy (kWh), capacity (kW), transmission rights, and financial risk management. Participants include utilities, federal power agencies, merchant generators, wholesale marketers, ISOs, transmission companies, and financial services providers. These services are discussed in greater detail on the following pages.

WHOLESALE SERVICES		
Service Provided	**Service Provider**	**Service Consumer**
Energy (MWh)	Merchant generators Utilities Federal power agencies ISOs Wholesale marketers	Utilities ISOs Wholesale marketers Retail marketers Large end users
Capacity (MW)	Merchant generators Utilities Federal power agencies ISOs Wholesale marketers	Utilities ISOs Wholesale marketers Retail marketers
Transmission Rights	Utilities Transmission companies ISOs	Merchant generators Utilities Wholesale marketers Retail marketers
Financial Risk Management	Financial companies ISOs Wholesale marketers	Merchant generators Utilities Wholesale marketers Retail marketers Large end users

SECTION THIRTEEN: MARKET DYNAMICS

Energy and Generation Capacity

Energy sales refer to the sale of electricity that will be generated and transmitted on the grid to the point of sale. In the wholesale marketplace this is measured in MWh. Capacity sales refer to the sale of a right to call on generation capacity if needed. Capacity is measured in MW. While energy sales are generally priced simply in $/MWh, capacity sales commonly have a two-part price, a $/MW that is paid for the reservation of capacity whether it is used or not, and an additional $/MWh that is paid if the unit is dispatched.

Two markets exist for both energy and capacity. Forward markets (which are commitments for a time in the future), and spot markets (which are commitments to deliver energy or capacity on the same day or for the day following the transactions). Within each category there are numerous types of contracts that are common.

Forward Markets

Common contract structures for forward markets include:

- Full requirements agreement — A full requirements agreement obligates the seller to provide all the energy and capacity required for the buyer. For a small utility, a full requirements agreement would replace the need to own its own generation. Common users of full requirements agreements are small municipal or co-op utilities that contract with a public power agency or with a federal power authority. Full requirements agreements may also be used by end-use customers who wish to contract with a marketer to provide for all their electricity needs.

- Partial requirements agreement – A partial requirements agreement obligates the seller to provide a fixed amount of capacity (MW) and/or energy (MWh) to the buyer.

- Firm power — This means that the seller has provided a fixed amount of energy and has reserved transmission to deliver the power to the buyer. Firm power usually must be paid for whether or not the buyer takes it.

- Life-of-the-plant — This is an agreement between a merchant power plant and a utility where the utility contracts for the capacity of the unit for as long as the unit is in service. Under this contract the utility would normally have the rights to dispatch the unit. This is used as an alternative to the utility building the unit itself.

- Pooling agreements — These are agreements among multiple owners of generation and buyers of energy that agree to put their assets into a pool and have them operated for overall optimization of the group.

- Tolling agreement or heat rate agreement — This is an agreement between a buyer and a power plant. The buyer agrees to provide the necessary fuel (usually natural gas) and to take the energy produced. The power plant is paid a fee for running the unit to convert the fuel into electricity. The plant must also agree to run the unit in such a manner as to meet a minimum heat rate and availability.

- Balance of the month — An agreement to provide energy for the remaining days in the current month.

- Non-firm power — This is energy that is sold on an as-available basis. There are no commitments as to availability of capacity and/or transmission and no commitment on the part of the buyer to purchase the power if he does not want it.

- Peak power — An agreement to buy/sell energy for the peak period. Unlike elsewhere in the industry, the wholesale market uses peak to refer to the 16 hours between 7 a.m. and 11 p.m. Much of the country trades 5 x 16 blocks, meaning the 5 weekdays and 16 hours per day. The western regions often trade 6 x 16 blocks which add in Saturday.

- Off-peak power — An agreement to buy/sell energy for the hours not defined as peak (see above).

Spot Markets

Spot markets generally refer to day-ahead sales (energy or capacity being sold for the following day) or sales for energy or capacity to be used on the day of the sale. Sales for the same day are commonly termed hour-ahead (for the next full hour) or imbalance energy. Imbalance energy (also called balancing energy) is energy bought or sold by the system operator to keep the system in balance within a specific hour. Day-ahead and hour-ahead spot market sales may be bilateral transactions that take place between two private market participants or they may be transactions that occur through an ISO (whereby the ISO is facilitating the buying and selling).

Transmission Rights

Transmission rights are used by generators and/or marketers to deliver power to a point of sale. Given the current state of the transmission infrastructure, rights to use transmission can make or break any given transaction. The way that transmission rights are made available varies depending on the market structure and trading arrangements (see Section Nine). Under the wheeling method, types of rights include long-term firm transmission (usually held only by utilities and under long-term contracts), wheeling firm transmission (firm transmission made available on a transaction-by-transaction basis under the utility open access tariffs), and wheeling non-firm transmission. Once a market has transformed to an ISO marketplace, transmission rights transition to a transmission congestion model. Under this model, transmission is allocated on an hour-by-hour basis in the spot market with congestion charges applying to users of any congested paths. To hedge against the financial uncertainty associated with congestion charges, ISOs often offer financial transmission rights (FTRs). FTRs are auctioned off to the marketplace and provide a fixed price guarantee associated with use of a certain transmission path.

Financial Services

Generators, marketers, and utilities holding assets or contracts and/or needing to purchase future supply that is subject to market fluctuations may turn to financial markets to hedge some or all of the price risk. Common products include price swaps (exchanging variable price risk for a fixed price) and options (used to create price floors and ceilings that reduce risk of price fluctuations). We will take a closer look at these concepts in Section Fourteen.

The Retail Market

Retail markets refer to transactions between the supplier and the end user of electricity services. Retail markets are split between those where retail access is permitted by regulation and those where utilities remain monopoly suppliers. Unlike the wholesale market which is extremely sensitive to price considerations, portions of the retail market are often more sensitive to service and relationships than to price. Many smaller end-use customers see electricity as a fundamental necessity for their homes and businesses, but cannot afford to focus too much on day-to-day transactions. For this reason, they are more likely to pay a premium to receive good service. A recent McKinsey Company survey of smaller customers in the European marketplace – which has seen a high level of retail competition – found that a majority of customers were turning away from the lowest-cost options, feeling they had been burned by the substandard service from these providers. The majority of customers were more attracted to low-hassle and technology-based services.

In the U.S., large consumers of electricity are always the first customers calling for regulatory reforms to allow them to choose their suppliers and generally quickly take advantage of supply options once they are available. Smaller customers as a whole are much more reluctant to switch suppliers given their uncertainty about the level of benefits associated with switching. The amount of load served by competitive suppliers has grown slowly but steadily in recent years in the U.S. In mid-2001, approximately 25,000 MW was served by competitive suppliers. By 2007, this had increased to over 70,000 MW, which is close to 10% of total U.S. loads. As markets mature in states with retail choice and regulatory price caps expire in some states exposing utility customers to market prices, it is expected that the amount of load served by non-utility suppliers will continue to grow.

Services in the retail market center on the provision of energy (kWh) and other related services. Other important services include energy efficiency, demand side management, power reliability, power quality, and energy information. These may be provided by a utility or any number of competitive suppliers. These services are discussed in greater detail below.

RETAIL SERVICES		
Service Provided	**Service Provider**	**Service Consumer**
Supply	Utilities Wholesale marketers Retail marketers	End-use customers
Value-added services	Utilities Retail marketers ESCOs	End-use customers

Utility Retail Services

Retail services provided by utilities are defined in the utility's tariffs and must be approved by the regulatory commission. Services are associated with specific customer classes such as residential, small commercial, large commercial, and industrial. All customers take utility distribution services whether or not they have access to competitive suppliers. Larger customers tend to have two-part rates that include a demand charge and an energy charge while smaller customers usually pay only energy charges. Most customers pay a monthly customer charge, which is a fixed amount per account based on the customer class. Energy and demand charges may be fixed throughout the year, or may vary by season (summer and winter rates are common). For larger customers, time-of-use rates are often applied where rates are different at different times of the day (in the summer, for instance, three rates might apply – a peak rate from noon to 6 p.m., a mid-peak rate from 8 a.m. to noon and from 6 p.m. to 11 p.m., with an off-peak rate applying to the remaining hours). Larger customers often have the option of choosing curtailable rates, real-time prices (which vary based on the market price) or taking service at higher voltage (primary or transmission level service). Utilities may also offer value-added services centered on energy efficiency or demand side management incentives which are designed to reduce overall peak loads and thus hold down costs for all customers. Utilities may also offer assistance with power reliability or power quality issues.

> **TYPICAL UTILITY SERVICES**
>
> - Residential
> - Residential Experimental Time-of-Use (TOU)
> - Residential Multi-unit
> - Small Commercial
> - Commercial Demand Metered
> - Commercial Demand Metered TOU
> - Industrial TOU
> - Industrial TOU Curtailable
> - Agriculture
> - Street Lighting

Competitive Retail Services

In regions where direct access is permitted, customers have the option of acquiring supply from competitive suppliers known as retail marketers (large customers sometimes buy from suppliers that focus mainly on wholesale trading since these users buy in large volumes, so you may also hear it said they are buying from wholesale marketers). For larger customers, offerings generally center on price and/or integrated services. In today's market where there are large price uncertainties, a typical pricing methodology is to offer specific kW blocks of power at a fixed price, with additional usage priced at a market index rate. In some cases suppliers may offer ceilings and floors for the portions priced at market rates. Integrated services offerings tie energy efficiency, demand side management, and/or multiple fuel services into one package and are attractive to customers who are not large enough to have in-house energy expertise and larger customers who choose to outsource this function. Small commercial and residential customers are more apt to buy full requirements power (power covering all their needs) at fixed prices which are adjusted periodically to account for market changes. These customers are generally not accustomed to buying additional services along with their power.

ESCO and Other Energy Services

Energy services companies, known as ESCOs, provide important services in either regulated or competitive retail markets. ESCOs work directly with end users to help them minimize their energy costs. Services include bill evaluation, energy efficiency, demand side management, usage monitoring, appliance maintenance, distributed generation, and in competitive markets, choice of suppliers and supply contract negotiation assistance. An additional important service offered is assistance with power reliability or power quality needs. This may come from an ESCO or an engineering or equipment firm.

What you will learn:

- How various market participants create profits
- How profits are created under traditional and incentive ratemaking
- Key skills for creating profits
- What risk management is and why it's important
- How market participants manage risk using physical and financial instruments
- The difference between hedging and speculating
- How Value at Risk (VAR) is used to measure risk levels

SECTION FOURTEEN: MAKING MONEY & MANAGING RISK

It goes without saying that the ultimate goal for all market participants is to make money. But since large portions of the electric industry continue to be dominated by regulated monopolies, the basic concepts that apply to making money do not necessarily apply to all the entities we have studied. Nor is there always a strong incentive to develop products and services solely focused on customer desires (since much of the ability to make money for a regulated entity is determined by regulators, not customers). The electric industry is further complicated by the unique mixture of regulated and non-regulated entities, as well as the variation of regulation and market structures from state-to-state and region-to-region. Thus it is critically important to understand the differing profit motivations of various market participants and how each makes money.

KEY SKILLS FOR PROFITABLE BUSINESSES		
Non-Regulated	**Traditional Regulation**	**Incentive Regulation**
• Marketing/pricing • Asset management • Financial management • Customer service • Billing • Credit and collections • Efficient operations • Information technology	• Regulatory/government relations • Expense containment • Asset expansion • Service reliability	• Purchasing • Expense containment • Productivity enhancement • Marketing/pricing • Information technology • Achieving service standards • Asset management

As we study the various ways in which market participants make money, we must also consider the inherent risks involved at all levels of the business. When we talk about risk, we mean the possibility that earnings will be lower than projected or lower than the market will support at the time products or services are delivered. Ten years ago, a section on risk in a book on the electric industry would have been very short. With all aspects of the industry regulated, the biggest risk a utility faced was regulatory risk – the risk that regulators would lower its rate of return or otherwise rule in such a way that made it difficult for the utility to earn an acceptable rate of return. Today we can't open a newspaper without seeing the extent to which industry players are at risk. In this section we will take a look at the various risks faced by electric market participants and the tools they employ to mitigate them.

How Market Participants Create Profits

In the simplest sense, a company's profits are the difference between revenues and expenses. Revenues are determined by the amount of products or services sold multiplied by the price that is charged. In a non-regulated environment, businesses usually attempt to set their prices so that earnings are maximized. This is determined by an analysis of the business' competition and the profits that can be attained at various pricing levels. The optimal price, however, is determined by market forces. Almost all utility prices – or rates – are determined based on the cost of providing services and not on market conditions.

WAYS TO MAKE PROFITS		
Non-Regulated	**Traditional Regulation**	**Incentive Regulation**
• Revenues exceed costs	• Increase authorized rate base • Reduce expenses • Increase authorized return	• Increase revenues • Reduce expenses/capital • Produce/buy below baseline • Achieve service standards

Under traditional cost-of-service regulation, a revenue requirement is determined so that all of a utility's costs – including a return on investment (profit) – are covered. Balancing accounts are used to ensure that utilities collect no more and no less than the approved revenues for certain items. The majority of a utility's earnings result from the return portion of the revenue requirement. In recent years, regulators have begun to experiment with incentive ratemaking in which the return is dependent upon the utility's performance. Clearly, the different ways of making money make for very different business models and corporate behaviors.

How a Utility Makes Money – Traditional Method

Using traditional methodology, the majority of earnings are determined by the rate of return on equity authorized by the commission multiplied by the cost of facilities in the rate base. Because earnings are dependent upon the value of the rate base, utilities have been incited to invest in more facilities, thereby increasing their potential earnings. To protect the customer from paying for unnecessary facilities, regulators require prior approval of major facility additions as well as reasonableness reviews of expenditures.

Under traditional regulation, a utility can increase earnings in four ways: increase rate base, increase the commission-approved rate of return on rate base, and in some cases, hold expenses below the forecast used to set rates and/or increase sales beyond forecasted loads. The latter strategies work if expenses and/or revenues are not subject to balancing account protection. So if a utility is able to get regulators to approve a certain expense threshold, and then subsequently beats that threshold, it gets to keep as profit any remaining expense dollars. Various regulatory authorities have different ways of treating expenses and revenues, so sometimes these opportunities are available to utilities and sometimes they are not.

It is important to keep in mind that rates of return are not guaranteed by regulators. Utilities can fall short of authorized returns in many ways. One area of risk is called disallowances. If regulators believe that utilities have failed to act prudently in spending money, the money spent can be disallowed, meaning that the utility is not allowed to include the expenditures in its revenue requirement. A second area of risk is expenses that are not balancing account protected. In this case, exceeding forecasted expenses would result in lower returns. Lastly, some utilities do not have balancing account protection for revenue fluctuations for weather. This means that should the utility's service area experience a cooler than usual summer, resulting in lower electric usage for air conditioning, revenues will be reduced. If there is no balancing account protection then the utility will fall short of earnings projections (conversely, if it is hotter than projected, earnings may increase).

One way of reducing utility risk is a mechanism called decoupling. Since profits for an investor-owned utility are determined by the amount of money left over after all expenses and debt have been paid, most utilities have more profit when they sell more electricity and less when sales fall. This can result in fluctuating profits due to weather which differs from forecast as noted above, and can also be a deterrent to utility support for energy efficiency measures. As a result, some regulators have implemented revenue decoupling, which removes the link between utility sales and utility profits. Under revenue decoupling, pre-defined over-collection or under-collection of revenues are recorded in a balancing account and applied to future year rates. Revenue decoupling makes utility earnings streams less risky and can encourage utilities to implement energy efficiency programs.

How a Utility Makes Money – Incentive Regulation

As the electric business restructures, regulators are likely to look at new ways of reforming the regulatory process. In some areas, traditional cost-of-service regulation is being supplemented with incentive regulation, which creates shareholder incentives for utilities to lower costs and reduce rates. As market-based rates become common in the wholesale marketplace, it will be harder for a utility to prosper under traditional regulation. Any time market rates are lower than the utility's cost of energy, intervenors will claim the utility performed unreasonably and that shareholders should bear some of the excess costs. Thus many utilities may prefer to accept the risks and rewards of incentive regulation. Under incentive regulation, utilities can increase profits by achieving or exceeding standards or market-based targets set by the regulator. For more information on incentive regulation, please see the discussion in Section Ten.

How Unregulated Market Participants Make Money

Unlike regulated entities, other market participants' profitability is driven by the harsh realities of market dynamics. These include whether the participant is selling a service that the market is wiling to buy, whether the participant is able to provide that service at a cost that still provides a reasonable profit given the price the market is willing to pay, and whether the participant is able to deliver the service after the product has been sold. In a volatile electricity marketplace, many entities have discovered that shareholder losses and even bankruptcy can be a heartbeat away! Especially hard hit have been the merchant generation companies who, at least in the short term, are finding that market electricity prices do not support the cost of building and running many units recently constructed. As the electricity business matures, strategies for profitability have begun to resemble strategies used by other competitive industries such as airlines and consumer product marketing.

Risk Management

Recent events in the electricity business make it very clear that no matter how a market participant makes money, the levels of risk encountered in the marketplace are so extreme that companies can go from apparent profitability to insolvency in the course of a few short months. All market participants must actively and thoroughly manage their risks at all times. If they aren't doing this, they are placing their shareholders' investment in severe jeopardy.

As we have seen, electricity prices have proven to be highly volatile. The graph on the right shows peak prices for three different pricing points (Cinergy Hub in Ohio, SP15 in southern California, and ERCOT in Texas) for the summer of 2006. As you can see, prices during these four months varied dramatically. For example, prices at Cinergy Hub ranged from a low of $33/MWh on June 7 and September 1 to a high of $144/MWh on August 1 and 2. This is more than four times the low price. Either of these extremes can mean a huge loss for an unprepared generator, marketer or end user who is at risk for such price fluctuations.

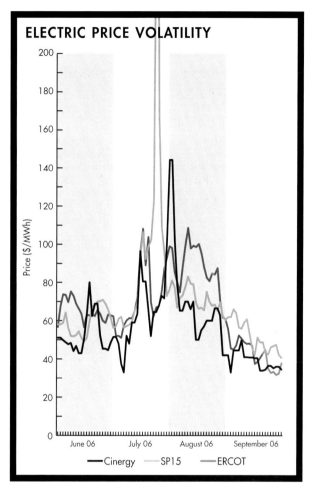

Because there are separate regional grids in the U.S., which often have transmission constraints within them, prices will vary significantly in different regions. And unlike the natural gas industry, where there is relatively strong correlation between prices in various regions, correlation between regions is often non-existent in the electric business.

The risk that we have been discussing so far is price risk. This is just one of many risks that are currently inherent in the electric marketplace:

- Price Risk — The risk that prices will move in the opposite direction to what the participant desires. This risk can be even more extreme for generation owners who not only face the risk of electric market pricing, but also the risk of price movement in their fuel supply.

- Volume risk — The risk that a customer does not use as much or uses more electricity than the supplier anticipated. This risk can result in unexpected spot purchases and/or sales or unexpected balancing charges.

- Transmission risk — The risk that a transmission path necessary to complete your transaction will be congested, resulting in unexpected transmission congestion charges or the need to buy replacement power at a higher price.

- Counterparty risk — The risk that any party you do business with will not honor its commitments.

- Execution risk — The risk that someone within your company fails to execute a transaction properly (e.g., a contract is not signed or a contractual condition not met).

- Tariff or regulatory risk — The risk that the regulator will change the rules applying to a business transaction after you have signed the contract.

- Operational risk — The risk that an asset you counted on fails to operate as expected.

Choices for Managing Risk

Risks are managed by developing and implementing a thorough internal risk management program, which includes measuring risk levels on a daily or hourly basis, structuring of physical transactions, use of financial instruments, and careful management of counterparty relationships. Following are descriptions of some of the most common risk management techniques used in the electric industry today.

Physical Risk Management

A physical deal is an agreement to buy or sell electric supply or transmission rights. Depending on the existing market position of the parties, physical risk management can include fixed pricing, pricing tied to market indexes, pricing with ceilings and floors, and selling physical blocks (meaning that the seller is providing a specific amount of kWh so that there is no risk of having to go out and buy more power on the spot market to supply the customer). Other means of physical risk management include setting up alternate contracts with other suppliers, having a right to call on generation capacity at a specific price, tolling agreements (whereby the buyer has the rights to use generation capacity at a specific price and is responsible for supplying the fuel), fixed pricing on fuel contracts, and building a portfolio of deals so that any single deal does not expose a large portion of a company's finances. One last means of physical risk management is owning assets yourself, including generation and fuel assets and/or signing long-term fixed price contracts to utilize assets owned by another party.

Most marketers or generators would prefer to make only transactions that are structured in such a way that profits are locked in before the contract is signed. The problem is that such a requirement can severely limit the number of transactions available. Another problem with physical deals is that they can be difficult to unwind (or get out of) if the market changes. Thus market participants often depend on a combination of physical deal structures and more liquid financial instruments to manage risk. Unfortunately, as we will learn next, the electricity industry is lacking in liquid financial instruments. In fact, physical sales are often more liquid (meaning they can be easily re-traded) than financial instruments in the electric industry.

Financial Risk Management

The use of financial instruments to manage risks has been common in commodity industries for decades. Financial instruments provide a means for market participants to shift risk without actually trading a physical commodity. Instruments used to do this include exchange-based futures and options, ISO-based financial transmission rights (FTRs) and over-the-counter (OTC) derivatives. Exchange-based instruments are specific defined contracts that trade on a central exchange like the New York Mercantile Exchange (NYMEX), while OTC instruments are simply contracts between two private parties.

Unlike the natural gas industry where the market for financial instruments has become very robust, trading volumes remain small in the electricity industry, thus limiting financial risk management opportunities. Reasons for this lack of liquidity include the lack of price correlation between regional hubs, the limited amount of power traded in many wholesale markets due to continued dominance of vertical utilities, and the wide variation of amounts of power used at different times during the day. There is currently a futures market run by NYMEX that trades futures and options for power in PJM. Both NYMEX and the Intercontinental Exchange (ICE) offer services that match buyers and sellers for various products at other hubs including futures-like instruments, swaps, location spreads, and options. Each ISO offers financial instruments that allow parties to hedge the risk of transmission congestion charges for specific transmission paths. These are commonly called FTRs (financial transmission rights), although different ISOs use different terms including transmission congestion contracts or TCCs (NY ISO) and transmission congestion rights or TCRs (ERCOT).

FINANCIAL INSTRUMENTS

Futures — A future is a supply contract between a buyer and seller where the buyer is obligated to take delivery and the seller is obligated to provide delivery of a fixed amount of commodity at a predetermined price and location at a specific time. Although futures can result in physical delivery, they are usually used as financial instruments by simply trading them back at or near the close of the trading period.

Option — An option is similar to a future, but differs in that it is the right, but not the obligation, to sell or purchase an amount of power.

OTC Instruments — Also known as derivatives, OTC instruments are more flexible than futures or options since they are not traded on an exchange, and can be designed as desired by the two parties involved.

Financial instruments are used by wholesale market participants to cover the risk of price fluctuations. For instance, a marketer may have a contract to sell a block of power to an end-use customer at a fixed price, but may be buying power at a market index. Rather than carry the risk of rising market prices, the marketer can lock in a margin by swapping the floating price (the indexed supply purchase) for a fixed price. Generators may also use instruments to lock in fixed revenues and are likely to be active in the fuel hedging markets to reduce their risks associated with coal, fuel oil and/or natural gas price fluctuations.

Speculation versus Hedging

To understand the use of financial instruments, you must clearly understand the difference between hedging and speculation. Hedgers reduce risk by paying a third party to assume that risk, much like a homeowner pays an insurance company to assume the risk of rebuilding her house in the event of a fire. Speculators, on the other hand, take on risk in the hopes of making money (for instance, if the insurance company takes in more money than it pays out in all of its fire claims, it has speculated successfully on the risk of its customers' fire losses).

On one side of a financial transaction, there is a party attempting to hedge risk. On the other side is a financial services company hoping to profit by taking on the risk of price volatility. This is achieved by charging a fee for the service, building a margin into any price guarantee, and/or designing a portfolio of transactions so the financial services company can profit by being a middleman between parties. For instance, if you were going to offer a product guaranteeing price, you might project the expected price level, add in $5/MWh to cover the risk, and add a couple more dollars per MWh for profit. It is critical for both sides of a transaction to carefully track what risk has been assigned to what party, and who is hedging versus speculating. Most of the negative stories about use of financial derivatives have occurred because firms were speculating and misjudged the level of risk, or because firms thought they were hedging but did not properly understand the level of risk to which they remained exposed.

Hedging Techniques

Risks can be hedged in a variety of ways. Following are examples of four common techniques used in the electricity business:

- Buying or selling at a fixed price — This requires no financial instruments as long as prices are fixed on both sides of a transaction. For instance, a marketer may agree to sell electricity to a Texas end user for one year at a price of $52/MWh. Because the price is fixed for the end user, that customer has no price risk for the

> ## A SIMPLE EXAMPLE OF HEDGING
>
> You are a small merchant generator with a contract to sell 10 MW of power to an end user for the month of April with a fixed price of $48/MWh. Your generator is fueled by natural gas and you are currently buying your gas supply at a floating monthly index price. Thus you are at risk that gas prices may increase to the point where you are unable to generate power at or below the fixed sales price. To hedge your risk, you might do the following:
>
> - Natural gas prices in your area for the current month are $5.50/Dth which, given the heat rate of your unit, is equivalent to a generation cost of $44/MWh. But you are unable to lock in a gas price for the month of April that you find attractive.
>
> - To cover your gas price risk, you buy NYMEX April futures to match the volume of gas you need to generate the 10 MW of power you have sold. NYMEX April futures are selling for $5.06/Dth.
>
> - When the end of the month approaches, you go into the market to buy gas for delivery to your power plant. You discover that gas prices have risen and you must pay $6.10/Dth to acquire the gas supply. You also see that NYMEX futures for April have gone up to $5.55/Dth. You buy the physical gas and sell the futures.
>
> - Your purchase of gas at $6.10/Dth results in your generating electricity at a cost of $48.80/MWh and selling it at $48.00/MWh. Not a good way to generate shareholder profits!
>
> - But with your hedging, you will sell your gas futures at a profit, thus covering your physical loss:
>
> | Date | Cash Deal | Futures Deal |
> |---|---|---|
> | March 7 | Sold electricity at $48.00/MWh | Bought gas futures at the equivalent cost of $40.48/MWh |
> | March 26 | Bought gas at the equivalent cost of $48.80/MWh | Sold gas futures at the equivalent price of $44.40/MWh |
> | Profit/Loss | ($0.80/MWh) loss | $3.92/MWh gain |
>
> After closing out all your transactions, your net gain on sale is $3.12/MWh. Your customer, by the way, is also pleased with your performance. Assuming natural gas generation is on the margin, it is likely electric market prices rose as well and your customer bought power at a below-market price.

length of the agreement. While the marketer has no price risk on the sale side of the transaction, she may be open to extreme risk is she hasn't locked in adequate electric supply at a specific price to cover the deal. Thus, she will attempt to find a generator willing to provide the supply at a fixed price.

- Using over-the-counter derivatives — Since pure exchange-based futures and options markets are limited to the PJM market (and then only to peak hours) most market participants must turn to OTC derivatives for financial risk management.

These instruments, offered by financial services companies, banks and some marketers mimic many of the features of the futures/options market but at different locations and under different terms. An example of an OTC derivative would be a price ceiling at the California-Oregon Border (COB). In this example, a bank guarantees a marketer that he will never pay more than $65/MWh for the summer months. If the price exceeds $65, the bank will compensate him for the difference between the higher price and the $65 ceiling. Another common OTC derivative is known as a price swap. Here someone holding an electric supply asset (either generation or a contract to buy electricity) subject to market price risk may "swap" the price risk to a financial services company and instead receive a fixed price. OTC derivatives can be extremely varied and the products offered can differ widely. Margins and transaction costs are often high since the financial services company is taking on substantial risk given the volatility of the electric industry.

- Using financial transmission rights (FTRs) – If a generator or a marketer must use a path that is sometimes congested to deliver power to a customer, the supplier is at risk for transmission congestion charges. These can quickly turn a profitable transaction into a loss. One way for the supplier to hedge this risk is to buy FTRs from the ISO. Each ISO has its own procedures for offering FTRs, but the general process is that financial rights associated with a specific path are auctioned periodically. The winner of the auction then has usage of that path for a guaranteed cost, regardless of the level of congestion charges. The supplier can then build this cost into his transaction pricing and is no longer exposed to this risk.

- Laying off counterparty risk – Over the last couple of years, a number of parties heavily involved in electric trading have either gone into bankruptcy or have become financially insolvent. This has resulted in contracts that have not been honored. Market participants now pay close attention to counterparty risk before entering into a transaction. Means of handling this risk include refusing to do business with parties that don't have a solid financial rating, putting provisions into the contract that allow for termination of the agreement if the counterparty fails to maintain defined standards of financial strength, requiring the party to put up a significant portion of the contract value in a letter of credit or escrow payment, and trading through a clearing exchange such as ICE or NYMEX that includes provisions to compensate parties for counterparty losses.

Value at Risk

Whatever means is being used to manage risk, it is critical for management of a company to actively measure the aggregate risk level it has incurred on at least a daily, if not an hourly, basis. The risk that is measured is the risk to the company's expected earnings stream if certain movements in market price or other detrimental events were to occur. This aggregate risk is measured by creating a "book" that shows all physical and financial positions, and using this information to estimate the earnings impact of various potential price movements. Procedures must also be in place to catch accidental execution mistakes or unauthorized actions of employees who may be trading outside of the guidelines given by management. We are all too familiar with the potential for huge impacts caused by failures in risk management.

A common way of measuring aggregate risk is called Value at Risk (VAR). In industry jargon, VAR can be described as "an estimate of a portfolio's potential for loss due to market movements, using standard statistical techniques and an estimate of future market volatility." In layman's terms, VAR is a calculation that attempts to assess how much total risk a company has taken on over any given period of time.

Unfortunately, VAR is only an imperfect means of quantifying actual risk. To calculate VAR, an assumption is made as to what level of market volatility will be experienced and then a level of statistical certainty is chosen (often 95%). Given a 95% certainty, your actual Value at Risk will theoretically exceed your calculation 18 days out of the year (5% of the time). And if one of those days is a day when electricity prices spike well above expected levels you can lose a lot of money quickly! In reality, one number cannot adequately reflect the complex risks encountered in today's marketplace. Thus it is always important to understand that the level of risks in the marketplace are inherently high, and that no means of analysis or use of risk management techniques can fully hedge all risks.

Despite this caveat, VAR is useful for a number of purposes:

- Quickly quantifying risk associated with a specific transaction.
- Comparing risk associated with expected return for alternative transactions.
- Quantifying risk across a portfolio of transactions (rather than looking at each transaction individually).
- Evaluating overall corporate risk profiles.
- Setting limits on allowed risk either by specific trader, specific business unit or corporate-wide.

As the industry becomes more familiar with the risks associated with the electricity business, new measures that go beyond VAR are being developed. It is likely that in the years to come, risk management will continue to evolve and become more sophisticated.

What you will learn:

- How the generation, transmission, distribution, system operations, and retail sales sectors may evolve

- A vision for a sustainable energy future

15

SECTION FIFTEEN: THE FUTURE OF THE ELECTRICITY BUSINESS

As we have seen in this book, the electricity industry has gone through radical change since the mid-1990s. Some regions, once dominated by vertical utilities that controlled all aspects of the business, have now become competitive markets with robust wholesale marketplaces and active competition for sales to end-use customers. Other regions remain dominated by strong vertical utilities and have felt little of the turbulent marketplace changes. Evolution of natural gas generation technologies led to a wave of construction of combined-cycle and combustion turbine natural gas units that have become a key part of the generation mix. And as we enter the end of the first decade of the new century, the issue of greenhouse gas emissions has become paramount, leading to rapid growth in renewables and the possibility that the future generation mix will significantly change. Regulation of greenhouse gas emissions either through expanded state programs or through federal legislation appears highly likely in most markets. Also changing the marketplace is renewed focus on demand side management and the inclusion of economic demand response as a supply resource in many markets.

The one thing we can be sure of is that the electric business will continue to see frequent change. While no one can predict a precise course of events, a look at ongoing trends and past experience in other businesses might give us some indication of what to expect. And it is worth noting that of all the countries in the world that have initiated electric deregulation, none have gone back to the former model of vertical regulation.

A Review of Market Changes

In the mid-1990s, all sectors of the industry were vertically integrated in monopoly utilities and virtually all pricing was based on a regulated cost-of-service model. This is still true in some regions of the U.S. But in other areas including the Northeast, portions of the Midwest, Texas, and portions of the West, wholesale competition is well established.

We have seen the beginnings of competitive generation in virtually all areas, first through power generated by QFs and then by merchant generators. Many regions of the country have turned power control functions over to ISOs which use a market-based approach to acquiring reliability services and allocating access to transmission. Some areas have seen the utilities divest of generation and in a few cases, even transmission. We are beginning to see transmission-only companies emerge, and these entities may build the next wave of transmission infrastructure. In 2009, close to 10% of electric load was acquiring supply through competitive suppliers rather than from the monopoly utility. Interestingly, although reports of market manipulation and energy company financial struggles dominated the headlines during this time of evolution, average retail rates fell steadily during most of the restructuring period, until rising in the last couple of years due to extremely high natural gas prices.

The Future of the Generation Sector

The merchant generation business has proven to be a volatile business. During the late 1990s and early 2000s a large amount of generation was built and much of it was financed with high levels of debt. Many areas of the country ended up with excess capacity thanks to an economic slowdown and a slowdown in restructuring which restricted opportunities for competitive generation in some areas. The result was many merchant generators found themselves struggling to stay in business. Restructuring of generation ownership and financing followed. In some areas this included utilities buying formerly competitive units to bring them into rate base, while in others new well-financed generation companies bought up assets at a significant discount to their cost of construction. As demands continued to grow and construction of power plants slowed, merchant generation in the mid-2000s again became a profitable business. But with the economic downturn in 2008 resulting in falling demand and lower market prices and tight credit, many merchant generators again struggled.

Interest remains high in alternative technologies. In recent years the vast majority of new generation built was fueled by natural gas. But the desire to diversify generation portfolios and reduce environmental impact has favored renewable generation development. Wind generation has become a favorite of many companies since it has no fuel price risk and minimal environmental risk. The expansion of wind power is likely to continue, although it will remain only a part of the overall mix given the need to match intermittent resources with other sources of supply. Interest is strong in finding cleaner ways to generate with coal including development of IGCC units and retrofit technologies that can limit carbon emissions from existing units. Given concerns over the effect of continued release of greenhouse gases into the atmosphere, interest has returned in new nuclear generation technologies and a handful of companies have begun the licensing process for new plants. And while not exactly generation, development of new electric storage technologies may finally break the paradigm that generation output must instantaneously match system demand.

In the longer term, the path for new generation is highly uncertain. Differing visions include growth of centralized renewable technologies such as wind farms, dominance by a new generation of coal and nuclear technologies, and the evolution to a decentralized generation sector dominated by fuel cells and other new distributed generation technologies.

The Future of Transmission

The current transmission grid is highly balkanized. Ownership (and in many cases regulation) of the grid is based on the historical evolution of monopoly utilities – not on market conditions. The transmission grid also appears in need of significant upgrading since construction of new facilities has stalled in recent decades. The current grid was built under the vertically-integrated utility model and has not been expanded to account for increased interstate wholesale trading, recent load growth or new construction of merchant generation plants. Continued evolution of the transmission sector is expected. New market participants focused on efficient operation and development of new transmission infrastructure are likely to become dominant over the next several years. Technological changes may allow existing transmission lines to carry more power and more effectively manage the flow through the grid, thus moving the U.S. in the direction of a robust transmission grid that is better suited to a competitive wholesale trading marketplace.

The Future of Distribution

The distribution sector will most likely remain a regulated monopoly for the foreseeable future. It is hard to imagine the evolution of competition for distribution services when this might entail two sets of distribution lines running in the same area. As current issues with deregulation dissipate and the benefits of competition become evident, it is likely that the model of distribution service separated from the supply function will become the standard – at least for industrial customers. Distribution companies will focus on efficient distribution operations under cost-of-service or incentive regulation. We may see consolidation of distribution companies as recent changes to federal laws have removed some barriers to utility mergers. From a logistical standpoint there is no reason to have over 3,000 distribution utilities in the U.S.

SECTION FIFTEEN: THE FUTURE OF THE ELECTRICITY BUSINESS

Growth prospects for increased distribution services include population increases in some regions, growth in power use for new appliances such as the various new electronic toys we fill our houses with, and development of new services enabled by smart grid technologies. The future of distribution may include a grid where customer appliances and meters communicate with monitoring devices on the distribution system. This could enable the grid to automatically respond to fluctuations or disturbances and to send pricing signals directly to customer appliances.

The Future of System Operations

For competition to work, it is clear that system operations must be run by an unbiased entity that has no position in the electric markets. While there is currently strong resistance to change in some regions of the U.S., over time the movement to ISOs and/or RTOs appears inevitable. And as benefits of competition become more transparent, it is likely that market participants will push for standardized market rules that allow participants to do business across the U.S. under similar terms. This may come through regulatory activities or it may come through voluntary market actions such as the many utilities that have chosen to join ISOs over the last few years.

In the long run, it is likely that we will have a common market structure run by a limited number of regional ISOs. These may evolve under the RTO model similar to today's ISOs or they may ultimately merge into the transmission function and be run by transcos in a manner similar to U.S. natural gas interstate pipelines.

The Future of Retail Marketing

Retail marketing is in its infancy in the U.S. This is one sector that will see immense changes in future years. As regulatory changes open up more of the marketplace, retail marketers will grow both in number and in market share. Companies with experience in other deregulated countries are quietly becoming active in the U.S. and Canadian markets, and some smaller regional marketers are growing at rapid rates. Coupled with technological changes that will open up the market for demand side management, home automation, real-time energy monitoring, and highly efficient and clean distributed generation, this evolution is likely to lead to completely new business models. The retail energy business may evolve to one based on services designed to sell specific value to customers. Customers may buy units of hot water, conditioned air, light, and appliance power from retail marketers who worry about how to efficiently create this value. Imagine a future where you buy hot water, comfortable room temperatures, light, appliance power, cable TV, on-demand movies, and high speed internet access all at one fixed price that is charged to your oil company credit card (which of course provides discounts on your gasoline purchases). Now that may get consumers excited about electric deregulation!

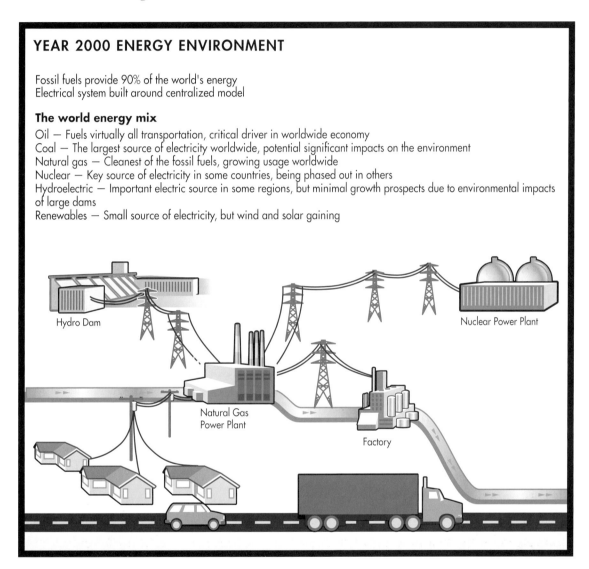

YEAR 2000 ENERGY ENVIRONMENT

Fossil fuels provide 90% of the world's energy
Electrical system built around centralized model

The world energy mix
Oil — Fuels virtually all transportation, critical driver in worldwide economy
Coal — The largest source of electricity worldwide, potential significant impacts on the environment
Natural gas — Cleanest of the fossil fuels, growing usage worldwide
Nuclear — Key source of electricity in some countries, being phased out in others
Hydroelectric — Important electric source in some regions, but minimal growth prospects due to environmental impacts of large dams
Renewables — Small source of electricity, but wind and solar gaining

A Sustainable Energy Future?

Most believe that eventually our society must transition to an energy future that is based on non-fossil fuels, although debate exists on whether that is in the near future or decades away. The key is to come up with sustainable sources of electricity that are friendly to the environment yet still cost effective.

SECTION FIFTEEN: THE FUTURE OF THE ELECTRICITY BUSINESS

YEAR 2050 ENERGY ENVIRONMENT

Fossil fuels no longer dominate
Renewables are important source of world's energy
Electrical system converted to distributed model with significant use of cogeneration
Hydrogen used as energy storage and transport medium
Efficiency of end-use devices has significantly increased

The world energy mix

Wind — Extensive network of wind farms becomes key source of electricity
Solar — Distributed network of solar cells provides electricity at users' locations
Natural gas — Still used for peaking electricity needs, remainder of resource devoted to hydrogen production
Nuclear — New breed of reactors may become important source of baseload electricity
Hydroelectric — Important electric source for peaking and system support, but environmental concerns and lack of undeveloped resources limit growth
Hydrogen — Hydrogen extracted from water or natural gas has replaced oil to fuel transport and has replaced natural gas as fuel for space and water heating.

One potential scenario (illustrated above) would replace today's centralized electric generation system with a more distributed system based on fuel cells and solar energy located at end-use locations. Automobiles would also be powered by fuel cells and might even act as home generators when parked in the garage. Wind and photovoltaic resources would contribute clean power and natural gas-fired generation may become a source used only for peaks. Hydrogen might be created from water using energy sources such as wind power or new generation nuclear. Hydrogen might also be created from natural gas at the wellhead and the current natural gas pipeline infrastructure could be upgraded to transport hydrogen. Hydrogen would then be piped to end users as a replacement for natural gas and fuel oil, and as a fuel for the ubiquitous fuel cells.

In the meantime, we must find a way to bridge the gap between today's world and the long-term future. Short-term changes will likely come through a combination of generation technology evolution and increased efficiency of electric use by the consumer. Technology changes that look promising include renewable generation, new processes for reducing emissions by fossil fuels, increased use of cogeneration, and continued improvements in end-use energy efficiency. In addition to the continued evolution of technology, changes in regulation and markets to reward reduced emissions and enhanced energy efficiency will be critical.

Electricity is evolving towards a new industry paradigm. As the industry evolves, there is no doubt that it will continue to be marked with revolutionary change. The future will be driven by the creativity and knowledge of individuals who expend the time to become experts not only in the electricity industry, but also in broader areas of sustainable development and servicing customer needs through continual innovation.

A

APPENDIX A: GLOSSARY

Ancillary services — The services in addition to electric supply that are required to deliver electricity to end users and to maintain system reliability. These include automatic generation control, reserves, voltage support, and black start.

Balancing power — See imbalance energy.

Baseload — Electricity usage that is constant through a specified time period. Also used to refer to the generating units that run all 24 hours of the day to serve a system's baseload demand.

Bilateral contract — An agreement between two private parties.

Blackout — The loss of power to a portion of the distribution or transmission system.

Black start — Generation that can start-up without energy from the grid.

Capacity — The maximum electric power output of a generating unit (measured in MW) or the maximum amount of power that lines or equipment can safely carry. Also generation that is available to ensure reliability.

Capacity payment — A payment for making generation capacity available to another party.

Carbon sequestration — The capture and storage of carbon as an approach to reducing greenhouse gas emissions during the power generation process.

Certificate case — Regulatory proceeding held to approve or deny construction of new facilities as requested by utilities.

Circuit — A complete path through which electricity travels, comprised of sources of electrons, energy consuming devices and conductors.

Circuit breaker — A device that interrupts electricity flow to a circuit by isolating the circuit from the source of electricity.

Cogeneration — The use of fuel to produce electricity as well as another product such as steam or hot water.

Combined-cycle turbine — A technology for generation that uses a fuel to drive two types of turbines in succession – a combustion turbine and a steam turbine.

APPENDIX A: GLOSSARY

Commercial customer — An end user that uses power to create a service. Sometimes also used by utilities to refer to manufacturing customers smaller than 500 kW.

Commodity — A standardized product or service that is easily traded among market participants. Also used to refer to electric supply.

Complaint case — A regulatory proceeding held to evaluate a complaint that a utility failed to properly follow regulatory rules.

Conductor — A material through which electricity can flow.

Congestion — A condition that occurs when the amount of requested transactions across a transmission path exceeds the physical capacity of that path.

Congestion charge — A charge by an ISO to market participants who utilize a congested path.

Congestion management — The process of allocating transmission capacity when congestion occurs.

Control area operator — The entity that performs system operations in a specific region, also called system operator.

Cost of service — The total amount of money, including return on invested capital, operation and maintenance costs, administrative costs, taxes, and depreciation expense required to provide a utility service.

Cost-of-service regulation — A regulatory methodology that allows utilities to charge rates designed to collect revenues equivalent to their cost of service

Counterparty — One of the participants in a financial contract.

Creditworthiness — An evaluation of a customer's or trading partner's financial accountability.

Current — The rate of flow of electrons through a conductor.

Customer choice — The ability of an end-use customer to choose his electricity supplier.

Customer charge — A fixed amount paid by a customer regardless of actual demand or electric consumption.

Customer class — A group of end users with similar characteristics, used to segment customers for the purpose of setting rates.

Demand — the total amount of electricity used at any given moment in time, usually measured in kW or MW.

Demand charge — The portion of a rate that is based on the maximum demand recorded over a specified period of time.

Demand curve — A graph showing demand plotted across time.

Demand side management — The act of reducing energy use or moving energy use from peak to off-peak periods to reduce overall energy costs.

Deregulation — The process of decreasing or eliminating government regulatory control over industries and allowing competitive forces to drive the market.

Distributed generation — Generation located at an end-use customer's facility.

Distribution — The delivery of electricity over medium and low-voltage lines to consumers of the electricity.

Divestiture — The selling of assets by a regulated utility as part of deregulation.

Economic demand response — Programs that offer end-use customers the opportunity to modify their electric usage in response to wholesale market price signals.

Electric co-op — See rural electric co-op.

Electricity — The flow of electrons through a conductor.

End user — The ultimate consumer of electricity.

Energy efficiency — The act of using less electricity to perform the same amount of work or to get the same end value.

Energy services company (ESCO) — A company that provides services to end users relating to their energy usage. Common services include energy efficiency and demand side management.

Fault — A failure or interruption in an electrical circuit.

Federal Energy Regulatory Commission (FERC) — the federal body that regulates wholesale electric services.

Federal power agency — An agency of the U.S. government that markets the output of generating units owned by the federal government.

Financial services company — An entity that provides risk management and financing services.

Firm service — Supply or transmission service that is expected to always be available except during operational problems.

APPENDIX A: GLOSSARY

Financial transmission right (FTR) — A right to receive financial compensation for the difference between actual congestion charges and the price of the FTR.

Forward market — A market where delivery of the item purchased is at some future point in time.

Frequency — How often the direction of flow reverses in an AC circuit.

Fuel cell — A device that converts stored chemical energy directly to electric energy. Although similar to a battery, the major difference is that a fuel cell operates with a continuous supply of fuel (such as natural gas or hydrogen) as opposed to a battery which contains a fixed supply of fuel.

Futures contract — A supply contract between a buyer and seller where the buyer is obligated to take delivery and the seller is obligated to provide delivery of a fixed amount of commodity at a predetermined price and location at a specific period in time. Futures are bought and sold through an exchange such as NYMEX.

Generation — The creation of electricity.

Generator — Used synonymously with the term power plant (although technically, the generator is the part of the power plant that converts the mechanical power of a spinning shaft to electricity).

Global warming — The warming of the earth's atmosphere due to increased concentrations of greenhouse gases.

Green power — Electricity generated using renewable fuels, usually excluding large hydro power.

Grid — Usually used to describe the interconnected transmission system, although sometimes used with distribution (distribution grid) to describe the distribution system.

Heat rate — The amount of fuel required to generate a specified amount of electricity, usually expressed in terms of Btu/kWh.

Hedge — The initiation of a transaction in a physical or financial market to reduce risk.

Hub — A physical location where multiple transmission lines interconnect and where buyers and sellers can make transactions.

Hydro power — Electricity generated by water falling across a water turbine.

Imbalance — The discrepancy between the amount of electricity an entity delivers into the grid and the actual amount the entity consumes.

Imbalance energy — Power purchased by the system operator during the hour to keep the system supply in balance with demand.

Incentive ratemaking — A form of ratemaking that rewards utility shareholders for achieving goals set by the regulator.

Independent power producer (IPP) — A generation company that is not part of a regulated vertically-integrated utility company that sells output under a long-term contract.

Independent System Operator (ISO) — An independent entity that provides system operation functions including managing system reliability and transmission access.

Index — A calculated number designed to represent the average price of electricity bought and sold at a specific location during a specified period of time.

Industrial customer — An end user that uses power for manufacturing or production of a product. Sometimes defined by utilities simply by size – bigger than 500 kW demand is a common minimum size.

Insulator — A material with high resistance to electricity, meaning that electricity cannot easily travel through it.

Integrated resource planning — The process by which a utility forecasts future demand, evaluates all its options for satisfying that demand and then develops a plan for serving it.

Interruptible rates — An electric rate schedule whereby the end-use customer agrees to not use power during certain hours when instructed by the system operator (used by the system operator as a means of maintaining reliability). In return, the customer receives a rate discount.

Investor-owned utility (IOU) — A regulated monopoly utility that is owned by shareholders and run as a for-profit entity.

kW — Kilowatt.

kWh — Kilowatt-hour.

Kyoto Protocol — An international treaty signed by most industrialized counties in 1997 in Kyoto, Japan and put into effect in 2005. The treaty has provisions for action by countries to reduce the emission of greenhouse gases.

Load — An amount of end-use demand.

APPENDIX A: GLOSSARY

Load serving entity (LSE) — An entity that sells electric supply to an end user.

Locational marginal pricing (LMP) — A method of setting prices in an ISO market whereby prices at specific locations on the grid are determined by the marginal price of generation of power available to that specific location. Prices vary from location-to-location based on transmission congestion and losses.

Loop flow — Flow of electricity that follows the path of least resistance on the transmission grid. The actual path may include parallel paths around the assumed contractual path.

Market-based rates — Charges for regulated services that are determined by market forces rather than being set by the regulator.

Marketer — An entity that buys electricity, arranges for its transmission and then resells the electricity to end users or other electricity buyers.

Market power — The ability of a market participant to artificially elevate prices over a period of time.

Merchant generator — A generation unit or company that is not part of a regulated monopoly vertically-integrated utility and that is subject to market pricing for sales.

Meter — A device used to measure the amount of electricity flowing through a point on the system.

Monopoly — A marketplace characterized by only one seller.

Muni — See municipal utility.

Municipal utility — A utility owned and operated by a municipality.

Native load — The end-use customer load of a specific utility.

Natural gas — A combustible gaseous mixture of simple hydrocarbon compounds, primarily methane.

New York Mercantile Exchange (NYMEX) — An organization that runs the market for trading of commodity futures and options.

Notice of Proposed Rulemaking — A document released by a regulatory agency in which the agency sets forth a proposed revision to its rules and gives market participants notice concerning the regulatory proceeding that will consider these revised rules.

Off-peak — The hours during the day when demand is at its lowest.

Open access — The requirement that a transmission system transmit electricity for any creditworthy party on a non-discriminatory basis.

Option — A contract that gives the holder the right, but not the obligation, to purchase or sell a commodity at a specific price within a specified time period in return for a premium payment

Overhead facilities — Electrical facilities that are installed on transmission towers or distribution poles.

Peak — The hours during the day when demand is at its highest.

Peak demand — The maximum demand for electricity in a given period of time

Peaking units — Generating units run only during times of peak demand on a system.

Performance-based ratemaking (PBR) — A form of incentive ratemaking in which a utility's actual performance (either financial or service-wise) is compared against specified baselines. The utility can attain extra earnings if the baseline is exceeded, but can lose earnings if the baseline is not achieved.

PJM — The the largest ISO in the U.S., which is the system operator for parts of Mid-Atlantic, Northeast and Midwest states.

Power — A synonym for electricity.

Power pool — An entity formed by multiple utilities to coordinate dispatch of generating units owned by the utilities to optimize coordinated system operations among the utilities.

Power quality — A measure of the level of voltage and/or frequency disturbances.

Power purchase agreement (PPA) — A contract for the sale/purchase of electricity.

Price volatility — The movement of market prices over time.

Public Service Commission (PSC) — The state agency that regulates the activities of investor-owned utilities (and also municipal utilities in some states).

Public Utility Commission (PUC) — See Public Service Commission

Public utility — A regulated entity that supplies the general public with an essential service such as electricity, natural gas, water, or telephone.

Public utility district (PUD) — A utility run by a local governmental agency or a group of governmental agencies, other than a municipality.

Rate — A regulated price charged by a regulated entity such as a utility.

APPENDIX A: GLOSSARY

Rate base — The net investment in facilities, equipment and other property a utility has constructed or purchased to provide utility services to its customers.

Rate case — The regulatory proceeding in which a utility's rates are determined.

Rate design — The development and structure of rates for regulated electric services.

Rate schedule — The commission-approved document setting out rates and terms of service specific to a certain service and service provider.

Regional transmission organization (RTO) — An ISO that operates over a regional geographical area and fits specific criteria defined by FERC.

Regulation — The multitude of rules or orders issued by state or federal agencies that dictate how electric service is provided to customers. Also, used in system operations to describe ramping a generating unit up or down to match supply to demand in real time.

Regulator — The governmental entity that sets the rules and orders that make up regulation.

Renewable fuel — A fuel that is naturally replenished such as wind or solar.

Reliability — A measure of how often electrical service is interrupted.

Residential customer — An end user that uses power in a home.

Reserves — Generation capacity that is available to the system operator if needed, but that is not currently generating electricity.

Restructuring — Changes in regulatory rules that result in change in control, ownership or regulatory mechanisms applicable to specific industry sectors.

Retail access — The opportunity for an end user to buy electric supply from someone other than his regulated utility distribution company.

Retail competition — The opportunity for multiple electric suppliers to compete to sell electric supply service to end-use customers.

Retail marketer — A firm that sells products and services directly to end users.

Resistance — A measure of the strength of impedance, which is a physical property that slows down the flow of electricity.

Revenue requirement — The revenues a utility must take in to cover its total estimated costs and allowed return.

Rulemaking — A regulatory proceeding held to establish new market rules.

Rules — Commission-approved general terms of service included in tariffs.

Rural electric co-op — A utility owned by its customers, that usually serves rural areas.

Scheduling — The process of determining which generating units will be generating or on reserve status for a specific hour. Also, the process of determining which requested transactions across a transmission line will be allowed to occur.

Service territory — The geographical area served by a utility.

Short circuit — An interruption in the flow of electricity due to an undesired conductor coming in contact with the electrical flow.

Smart grid — Transmission and distribution system integrating modern digital technologies to enhance monitoring, communication, control, and support systems.

Speculating — The initiation of a transaction in a physical or financial market with the goal of making a profit due to market movement.

Spot market — The short term market for electricity – usually refers to day-ahead, hour-ahead and real-time markets.

Stranded costs — Utility costs that result from assets acquired under prior regulatory rules that are in excess of the market value of those assets.

Substation — A facility containing switches, transformers and other equipment used to adjust voltages and monitor circuits.

Supply — Electricity available to the grid.

System operator — The entity that manages the transmission grid by dispatching generation and scheduling reserves and transmission.

System peak — The maximum load on an electrical system during a given period of time.

Tariff — All effective rate schedules for a utility, along with the general terms and conditions of service.

Transformer — A device used to change voltage. A step-up transformer increases the voltage while a step-down transformer decreases it.

Transco — The abbreviation for transmission company, a regulated entity that owns only transmission facilities.

Transmission — The transport of electricity over high voltage power lines from generators to the interconnection with the distribution system.

APPENDIX A: GLOSSARY

Trading arrangements — The set of rules that specify how the system operator will acquire the necessary services to maintain system reliability and will allocate transmission access.

Underground facilities — Electrical facilities that are installed below ground level.

Utility distribution company (UDC) — A regulated utility that provides distribution services to end users.

Value-added services — Services related to electrical supply that are in addition to supply itself.

Value at Risk (VAR) — A measure of potential earnings loss due to adverse market movements with a specified probability over a particular period of time.

Vertical integration — The ownership of all sectors of electric delivery (generation, transmission, system operations, and distribution) within one entity.

Volatility — See price volatility.

Volt — A unit of measure of voltage.

Voltage — The electrical pressure that moves electricity through conductors.

Wheeling — The transmission of power across a utility system on behalf of a marketer or generator.

Wholesale trading — The buying and selling of power between parties that are not ultimate end users.

B

APPENDIX B: UNITS AND CONVERSIONS

Mcf = thousand cubic feet

MMcf = million cubic feet

Btu = British thermal unit

MMBtu = million Btu

GJ = gigajoule (metric measure of energy)

Dth = decatherm

kW = kilowatt

kWh = kilowatt hour

MW = megawatt

MWh = megawatt hour

1 therm = 100,000 Btu

1 Dth = 10 therms

10 therms = 1 MMBtu

1,000,000 Btu = 1 MMBtu

1 Dth = 1 MMBtu

1000 Mcf = 1 MMcf

1000 MMcf = 1 Bcf

1 MMcf = 1,015 MMBtu*

1 GJ = 0.95 MMBtu

1000 kWh = 1 MWh

1000 kW = 1 MW

*This conversion varies with the energy content of the gas

c

APPENDIX C: ACRONYMS

A — Amp

AC — Alternating current

AGC — Automatic generation control

ALJ — Administrative Law Judge

BPA — Bonneville Power Administration

Btu — British thermal unit

CO_2 — Carbon dioxide

CHP — Combined heat and power

COB — California-Oregon border

CPUC — California Public Utilities Commission

DC — Direct current

DG — Distributed generation

DOE — U.S. Department of Energy

DSM — Demand side management

Dth — Decatherm

ECAR — East Central Area Reliability Coordination Agreement

EDR — Economic demand response

EIA — Energy Information Administration

EMF — Electromagnetic field

EPA — Environmental Protection Agency

EPMC — Equal proportionate marginal costs

APPENDIX C: ACRONYMS

ERCOT — Electric Reliability Council of Texas

ESCO — Energy services company

ESP — Energy services provider

EWG — Exempt wholesale generator

FERC — Federal Energy Regulatory Commission

FPA — Federal Power Act

FPC — Federal Power Commission

FRCC — Florida Reliability Coordinating Council

FTC — Federal Trade Commission

FTR — Financial transmission rights

GNP — Gross national product

GW — Gigawatt

GWh — Gigawatt-hour

HVAC — Heating ventilating and air conditioning

Hz — Hertz

ICE — Intercontinental Exchange

IGCC — Integrated gas combined-cycle

IOU — Investor-owned utility

IPP — Independent power producer

IRP — Integrated resource plan

ISO — Independent System Operator

kV — Kilovolt

kW — Kilowatt

kWh — Kilowatt-hour

LMP — Locational marginal pricing

LSE — Load serving entity

MAAC — Mid-Atlantic Area Council

MAIN — Mid-America Interconnected Network

MAPP — Mid-Continent Area Power Pool

MISO — Midwest ISO

MMBtu — Million British thermal units

Muni — Municipal utility

MW — Megawatt

MWh — Megawatt-hour

NERC — North American Electric Reliability Council

NIMBY — Not in my backyard

NOPR — Notice of Proposed Rulemaking

NOx — Nitrogen oxide

NPCC — Northeast Power Coordinating Council

NRC — Nuclear Regulatory Commission

NUG — Non-utility generator

NYMEX — New York Mercantile Exchange

O&M — Operations and maintenance

OAT — Open access transmission

OTC — Over-the-counter

PJM — Pennsylvania-New Jersey-Maryland

POLR — Provider of last resort

PSC — Public Service Commission

PUC — Public Utility Commission

PUHCA — Public Utilities Holding Company Act of 1935

PUD — Public utilities district

PURPA — Public Utilities Regulatory Policy Act of 1978

APPENDIX C: ACRONYMS

PV — Photovoltaic

QF — Qualifying facility

R&D — Research and development

REA — Rural electric agency

REC — Retail electric company

ROE — Return on equity

RPS — Renewable portfolio standard

RTO — Regional transmission operator

SCADA — Supervisory control and data acquisition

SEC — Securities and Exchange Commission

SERC — Southeastern Electric Reliability Council

SEPA — Southeastern Power Administration

SMD — Standard Market Design

SPP — Southwest Power Pool

SWPA — Southwestern Power Administration

SO_2 — Sulfur dioxide

TCC — Transmission congestion contract

TCR — Transmission congestion right

TLR — Transmission loading relief

TO — Transmission owner

TOU — Time of use

TVA — Tennessee Valley Authority

UDC — Utility distribution company

UPS — Uninterruptible power supply

V — Volt

VAR — Value at Risk; also volt-ampere reactive

WAPA — Western Area Power Administration

WECC — Western Electricity Coordinating Council

D

APPENDIX D: INDEX

AC power 6, 14
Amperes 9, 12
Ancillary services 78-80, 106
    Automatic generation control 79
    Black start 80
    Non-spinning reserves 80
    Spinning reserves 79-80
    Supplemental reserves 80
    Voltage support 80
Arab Oil Embargo 7
Arc light 6
Atomic Energy Act 114
Atomic Energy Agency 114
Automatic reclosers 65
Balancing account 120, 177-185
Bilateral agreements 105, 106
California crisis 107, 133, 149, 153-155, 166
California ISO 153
California Public Utilities Commission 154, 155
Carbon dioxide 43, 44, 45
Carbon sequestration 53
Cascading outages 81
Centralized bidding 164
Circuit 9
Circuit breakers 65
Clean Air Act 45, 114
Clearing price 106, 107
Competition
    Creating a competitive market 136-140
    Supply side 133-134
Conductor 9
Congestion 106, 138, 155
Congestion charge 109, 181
Constraints 108
Construction 59
Consumers 21-33
    Commercial 25-28
        Average costs 28
        Electricity use 26
        Monthly consumption 27
    Industrial 28-31
        Average costs 31
        Electricity use 29
        Monthly consumption 30
    Number of 22
    Residential 22
        Average costs 25
        Electricity use 23
        Monthly consumption 24
    Usage by 21
Consumption, per-capita 4
Control 13
Control area 75
Cost-of-service pricing 159-173
Current 12
Davis, Gray 154
DC power 6, 14

APPENDIX D: INDEX

Demand curves 32-33
Demand side 135
Demand side management 46, 136, 145, 160
Distribution 63-70
    Common voltages 63
    Costs 69
    Future of 189
    Loop feed 67
    Network system 67
    Operation and planning 68
    Ownership 69
    Physical characteristics 63-66
    Radial feed 67
    Service configurations 66
    Types of 67
Divestiture 137
Dynamo 6
Edison Electric Illuminating Company 6
Edison Electric Light Company 1
Edison, Thomas 1, 6
Electric business
    Key physical sectors 18
Electric delivery system 15
Electric Reliability Organization 75
Electric storage 53
Electricity
    Key physical properties 16
Electrode 11
Electrolyte 11
Electromagnetic induction 12
Electrons, sources of 13
Energy
    Chemical energy 10
    Electrical energy 10
    Electromagnetic energy 10
    Mechanical energy 10
    Nuclear energy 10
    Thermal energy 10
Energy Policy Act 144-146
Energy services companies 88-92
Enron 149, 166
Environmental Protection Agency 116
Exempt wholesale generation 145
Federal Power Act 114
Federal power agencies 87, 143-156
Federal Power Commission 114
FERC 75, 114, 115, 117, 149, 154, 155
FERC Order 2000 148-149
FERC Order 888 146-148
Financial services companies 91
Financial transmission rights 138, 170, 181, 184
Forward market 106
Franklin, Benjamin 5
Fuel cells 52
General Electric 6
Generation 35-53
    Baseload 48
    Capacity by type 35
    Capital costs 40
    Characteristics 37
    Characteristics comparison 42
    Coal 36-37
    Demand response as alternative 46
    Distributed generation 42-43
    Environmental concerns 43-44
    Fuel oil 40-41
    Future of 188-189
    Hydro 39-40
    Intermediate 48
    Levelized cost of 51
    Natural gas 38-39

Nuclear 38
  Output by type 36
  Ownership 49
  Peaking 48
  Renewable 41-42, 137
  Typical usage 47
  Value of 50
  Variable operating costs 38
Gilbert, William 5
Harrison, Benjamin 6
Heat rate 39
Hertz 14
Hydrogen 192
Independent power producers 49, 88, 144
Independent System Operator 58, 75, 101, 115, 190
  By size and date 147
  Creating an ISO 138-139
Indexes 164-165
Integrated Gas Combined Cycle 37, 188
Integrated resource planning 49, 50, 100, 160
Intercontinental Exchange 163, 181
Investor-owned utilities 85, 86, 97, 115
Kirchoff's law 6
Kyoto Protocol 44-46
Least-cost dispatch 78
Leyden jar 5
Load factor 24
Load serving entities 74, 75, 92, 106
Load shape 32-33, 47
  Hourly 32
  Monthly 33
Locational marginal pricing 109
Marginal costs 82, 108, 155
Market dynamics 159-173
Market evolution 129-133

  Commoditization 132
  Deregulation 131
  Regulation 130-131
  Value-added services 132-133
Market power 133, 135
Market structures 95-109
  Complete retail competition model 102-104
  Definition of 95-97
  Single buyer with competitive generation model 99-100
  Vertically-integrated monopoly utility model 97-98
  Wholesale/industrial competition model 100-102
Marketers 88-89, 90-91, 102, 103
Merchant generators 89
Meter 68
Monopolies 111-125, 137, 170
  Natural monopoly 111
Morse, Samuel 6
Municipal utilities 86, 87, 97, 115
Must-run generation 78, 137
Must-take generation 78, 137
National Civic Federation 113
National Electric Light Association 113
National Energy Policy Act 114
National Environmental Policy Act 114
Native load 106, 146
NERC 75, 82, 106
New York ISO 181
New York Mercantile Exchange 181
Non-utility generators 49
Northeast blackout 7
Nuclear Regulatory Commission 114
Ohm's law 6, 10
Ohms 10, 12

APPENDIX D: INDEX

Pacific Gas and Electric 148, 154
PJM 148, 181
Portfolio theory 51
Power capacitors 65
Power factor 65
Power flow model 78
Power pools 88, 108
Power systems
    Operational characteristics 73-74
Price volatility 165-166, 179
Pricing 163-165
Provider of last resort 103
Public power agencies 86, 88
Public Utilities Regulatory Policies Act 114, 144
Public utility districts 86, 97
Public Utility Holding Company Act 114
Qualifying facility 48, 137
Rate base 120, 177-185
Rate classes 21
Rate of return 120, 177-185
Regional system operator 74
Regional Transmission Organization 148-149, 190
Regulation 111-125
    Certificate cases 116
    Complaint cases 116
    Environmental 45
    Federal 113-115
    Goals of 112
    Incentive regulation 122-123, 178-185
        Benchmarking 123
        Performance-based 123
        Rate caps 123
        Service standards 123
    Rate cases 116, 119-122
    Rate design 121-122
    Regulatory process 116-119
        Draft decision 118
        Final decision 118
        Hearings 117
        Initial filing 116-117
        Preliminary procedures 117
        Review of decisions 118
        Tariffs 118-119
    Rulemakings 116
    State 112-113
    Who regulates what? 115-116
Reliability 139-140
Resistance 12
Restructuring 127-141
    History of 143-156
    In other countries 156
    State 150-155
Retail market 170-173
    Future of 190-191
Retail services 171
    Competitive retail services 172
    ESCO and other energy services 173
    Utility retail services 171-172
Revenue decoupling 121, 177
Revenue requirement 121, 177-185
Risk management 175-185
    Financial risk management 181-184
    Futures 181, 183
    Hedging 136, 182, 183
    Options 181, 183
    Over-the-counter derivatives 183, 184
    Physical risk management 180-181
    Speculation 182
Rolling blackouts 153
Rural electric co-ops 86-87, 97, 115

Rural Electrification Administration 86
SCADA systems 68
Scheduling coordinators 106, 107
Schwarzenegger, Arnold 154
Service drop 66
Service line 16
Smart grid 2, 190
Southern California Edison 148, 154
Southwest Power Pool 150
Standard Market Design 149, 159-173
Stranded costs 147
Supply and demand 80-82, 82-83, 160-162
    Current U.S. situation 162
    Long-term 162
    Short-term 161-162
Supply side 133
Switches 65
System operations 73-83
    Changing role of 82
    Forecasting and scheduling 77-78
        Demand forecasting 78
        Example of scheduling 79
    Future of 190
    Unbiased 134-135
    Who handles 74-77
Telegraph 6
Tesla, Nikola 6
Thales of Miletus 5
Three Mile Island 7
Trading arrangements 105-109
    Decentralized 106-107
    Functions of models summarized 108
    Integrated 107-109
    Wheeling 105-106, 115, 169
Transco 59, 89-90, 115, 190
Transformer 12

Step-up 15
Transmission 55-61
    Access to 134
    Capacity 56
    Construction 56
    Costs 58
    Current status of 60-61
    Future of 189
    Operation and planning 57-58
    Ownership 58-59
    Physical characteristics 56
Transmission loading relief procedures 60
Transmission owners 91
TVA 150
Units 3
Utility distribution companies 91-92, 115
Value at Risk 184-185
VARs 65
Voltage regulators 65
Volts 10, 12
Watts 13
Westinghouse, George 6
Wholesale market 166-170
Wholesale services 167
    Energy and generation capacity 168
    Financial services 170
    Transmission rights 169-170